SAMUEL JOHNSON: AN ANALYSIS

Also by Charles H. Hinnant

THOMAS HOBBES
THOMAS HOBBES: a Reference Guide
PURITY AND DEFILEMENT IN *GULLIVER'S TRAVELS*

Samuel Johnson: An Analysis

Charles H. Hinnant
Professor of English
University of Missouri

MACMILLAN
PRESS

First published 1988

Published by
THE MACMILLAN PRESS LTD
Houndmills, Basingstoke, Hampshire RG21 2XS
and London
Companies and representatives
throughout the world

Filmsetting by Vantage Photosetting Co. Ltd
Eastleigh and London
Printed in Hong Kong

British Library Cataloguing in Publication Data
Samuel Johnson: an analysis.
1.   Johnson, Samuel, 1709–1784 — Criticism
and interpretation
I.   Title
828'.609   PR3534
ISBN 0–333–44339–X

Contents

Preface

This book grew out of the conviction that Samuel Johnson's famous review of Soame Jenyns's *A Free Inquiry into the Nature and Origin of Evil* was not an isolated event in eighteenth-century thought, as has generally been supposed, but one that can be related to a specific intellectual tradition. This tradition originated in a controversy between Isaac Newton and Gottfried Wilhelm Leibniz over the existence of the vacuum. In this controversy, it is clear that Newton did not, like Leibniz, regard the plenum, the notion that the universe is completely full, as a theological necessity. Indeed, he was not merely content to defend the possibility of producing an artificial vacuum but also presented arguments for the existence of the void in the sense of 'non-being', holding that the amount of matter contained in the universe is actually so small that it could be compressed into a nutshell. Though Newton's reasoning may seem subtle and even paradoxical to modern readers, it impressed many thinkers in his own time including several whose works Johnson is known to have read and admired — Isaac Watts, Herman Boerhaave, and George Cheyne. Johnson's response to Newton's arguments in defence of the vacuum is much more complex than that of these figures, for it bears not so much on scientific questions as on the religious, cosmic, social and moral implications of the Newtonian position.

The Newtonian dimension of Johnson's thought is clearest, perhaps, in his critique of the principles of plenitude, hierarchy, and continuity in his review of Jenyns's *Inquiry*. These principles acquire significance in a Newtonian context, I will suggest, because they represent an extension of the doctrine of the plenum that Newton attacked. Though Johnson would have regarded as pretentious any attempt to apply Newtonian ideas in a programmatic fashion, an underlying purpose of the argument of his review is to reject the possibility that anything like pure plenitude could serve as the basis for a moral or political theory. Insofar as the principle of continuity is a necessary corollary of the plenum, Johnson's review also insists upon the impossibility of this principle. His arguments against plenitude and continuity, my book will argue, are the foundation upon which Johnson's well-known preoccupation with the vacuities of existence is based. Just as good is inextricably intertwined with evil, so

vii

plenitude is inevitably inhabited by vacuity, and the moralist, like the natural philosopher, must acknowledge this condition.

The language of some of my chapters may seem unduly abstruse to readers unfamiliar with Johnson's grasp of philosophical issues, but the central question here is the highly traditional one of the relation between Johnson's richly scientific and technical vocabulary and contemporary trends in natural philosophy. An examination of the source of this vocabulary in Johnson's response to the New Science or to the philosophy of John Locke has been undertaken by a number of scholars including William K. Wimsatt, Jr, John W. Wright, David W. Tarbet, Paul Alkon, Howard W. Weinbrot, Arieh Sachs, and Richard Schwartz. What my analysis seeks to do is to relate this vocabulary to a quite different intellectual tradition. The chapters in this book sketch the various ways this tradition is reflected in Johnson's major texts and suggest above all what we might call the resiliency of his thought; the liveliness, brilliance and resourcefulness with which it responds to the disturbing sense of negation and the void introduced into philosophy by the Newtonian cosmology.

There is no doubt a certain danger in seeking to assimilate the radically diverse themes of Johnson's major writings to the shape of any single pattern. None the less, there is in Newtonian and post-Newtonian thought an outlook which is consistent with Johnson's very distinctive way of looking at the world. This outlook can be found in Newton's theories of the void, of the play of forces and extrinsic causality, and of the wide dispersion of matter in time and space. Newton, perhaps alone of the major intellectual figures of the early eighteenth century, provided a philosophical stance which could easily be adapted to Johnson's intensely personal experience of conflict and suffering, of indolence and melancholy as central elements of human existence.

Though I have been reading and thinking about Johnson for a number of years, I did most of the writing of this book during a sabbatical leave of absence from the University of Missouri in 1985–86. My colleagues from the University of Missouri have been especially helpful in their encouragement and support. Catherine Neale Parke, William V. Holtz, and Allen Thiher read portions of this book and offered valuable suggestions for its improvement. Gilbert Youmans assisted me in developing my view of Johnson's theory of language, while Lois Shelton suggested ways in which the direction of the argument in the opening chapters might be clarified. John R. Roberts and Peter Markie were most patient in answering my various queries. To all of these, I am particularly grateful.

Abbreviations

Diaries, Prayers	*Samuel Johnson: Diaries, Prayers, and Annals*, eds E. L. McAdam, Jr and Donald and Mary Hyde (New Haven: Yale University Press, 1958).
Idler, Adventurer	*Samuel Johnson: The Idler and The Adventurer*, eds W. J. Bate, John M. Bullitt, and L. F. Powell (New Haven: Yale University Press, 1963).
Journey	*Samuel Johnson: a Journey to the Western Islands of Scotland*, ed. Mary Lascelles (New Haven: Yale University Press, 1971).
Letters	*The Letters of Samuel Johnson: with Mrs. Thrale's Genuine Letters to Him*, ed. R. W. Chapman (3 vols, Oxford: Clarendon Press, 1952).
Lives	*The Lives of the English Poets*, ed. G. B. Hill (3 vols, Oxford: Clarendon Press, 1905).
Poems	*Samuel Johnson: Poems*, ed. E. L. McAdam, Jr, with George Milne (New Haven: Yale University Press, 1964).
Rambler	*Samuel Johnson: The Rambler*, eds W. J. Bate and Albrecht B. Strauss (3 vols, New Haven: Yale University Press, 1969).
Rasselas	*Johnson: History of Rasselas, Prince of Abyssinia*, ed. G. B. Hill (Oxford: Clarendon Press, 1887).
Sermons	*Samuel Johnson: Sermons*, eds Jean Hagstrum and James Gray (New Haven: Yale University Press, 1978).
Vinerian Lectures	E. L. McAdam, Jr, *Samuel Johnson and English Law* (Syracuse University Press, 1951).
Works	*The Works of Samuel Johnson, L.L.D.* (9 vols.: Oxford: Talboys and Wheeler, 1825).

1

Johnson and Newton:
Dismantling the Plenum

In *Rambler*, No. 8, Samuel Johnson writes:

> It is said by modern philosophers, that not only the great globes of
> matter are thinly scattered thro' the universe, but the hardest bodies are
> so porous, that, if all matter were compressed to perfect solidity, it
> might be contained in a cube of a few feet.   (*Rambler*, I, 41)

Although Johnson's wide familiarity with what is loosely called the New
Science of the seventeenth and early eighteenth centuries has long been
recognized, his precise relation to that movement has not always been
clearly understood.[1] In the above passage, which has never been
annotated to my knowledge, Johnson is referring not to all 'modern
philosophers' but, rather more obliquely, to Sir Isaac Newton and his
followers. In the course of a long and bitter controversy with Leibniz, a
controversy that was conducted mainly through intermediaries, Newton
came to reject the Cartesian–Leibnizian doctrine of the plenum.
Contending that this doctrine raised insuperable difficulties, Newton
argued that corporeal bodies are much more rare and tenuous than is
commonly believed.[2]

This argument is what Johnson is clearly alluding to in *Rambler*, No. 8.
Here, as elsewhere in his writings, he is demonstrating his knowledge not
only of a broad and general tradition of scientific thought, but of a specific
current of ideas within that tradition. The immediate source of this
knowledge need not concern us here. What ought to be considered rather
is its significance for Johnson's writings. That is the question which this
and subsequent chapters explore from different perspectives and in the
various contexts of Johnson's major works. In them I will argue that
Johnson was sufficiently familiar with Newton's arguments concerning
matter and the void that they could be viewed as the starting point for his
meditations on a wide range of subjects. Their influence is most evident in
the particular kind of logical reasoning Johnson employed in his review of
Soame Jenyns's *A Free Inquiry into the Nature and Origin of Evil*. The

parallel between Johnson's procedure and that of Newton will, if my analysis is valid, become apparent before the end of this chapter.

In coming to an understanding of how Newton's arguments might serve as a context for Johnson's reasoning in his review of Jenyns's *Inquiry*, we might begin by observing that Newton's critique of the plenum represents much more than the relinquishing of a venerable relic of a simpler but, alas, vanished past. On the contrary, it is involved in difficulties every bit as daunting as the position it attacks. In repudiating the notion of an absolutely full universe, Newton presupposes the existence not only of an external, three-dimensional void space, but of a space that is internalized within matter, rendering it porous and divisible. Yet inasmuch as anything that is porous is divisible without limit – it is always possible to conceive of half of anything that has any magnitude at all – so matter that is porous must be infinitely divisible. In the second and subsequent editions of *The Opticks*, Newton explores the conundrums raised by this issue, holding that if we 'conceive' of bodies as being composed of 'Particles . . . so disposed amongst themselves, that the Intervals or empty Spaces between them may be equal in magnitude to them all', then we may also conceive of these particles, in turn, as being 'composed of other Particles much smaller, which have as much empty Space between them as equall all the Magnitudes of these smaller Particles'. Because the particles have dimension and are, therefore, infinitely divisible, the presence of an interstitial vacuum between particles is conceivable, Newton suggests, only insofar as every particle is composed of smaller particles divided by intervals or empty spaces. This *regressus* can be halted, Newton implies, only by the postulation of invisible, infinitely small particles, 'such as have no Pores or empty Spaces between them'.[3] Naturally, this ultimate reduction means going beyond every sensible and factual limit. It implies an ideal limit comprehending an infinitely small body of matter, not an empirical limit involving a finite, observable body of matter.

This is one of Newton's arguments purporting to demonstrate the impossibility of the plenum, but what it illustrates more convincingly are the difficulties attendant upon any system based on the concept of infinity. In place of a universe that is absolutely 'full', the application of this concept now leads to the postulation of a universe that is virtually 'empty'. Yet if Newton himself was acutely aware of this dilemma, it quickly became obscured in the much simpler and more dogmatic formulations of Newton's popularizers. In *A View of Sir Isaac Newton's Philosophy* (1728), Henry Pemberton confidently insists that 'this whole globe of earth, nay all the known bodies in the universe together, as far as we know, may be compounded of no greater a portion of solid matter, than might be reduced into a globe of one inch only in diameter, or even less'. And in *An Account of Sir Isaac Newton's Philosophical Discoveries* (1748), Colin Maclaurin holds that 'there actually is a *vacuum*; and that, instead of an infinite, necessary, and indivisible plenitude, matter appears to occupy but a very small portion of space, and to have its parts actually divided and separated from each other'.[4]

The conviction that the universe is a vacuum rather than a plenum extended beyond Newton and his immediate followers. Among writers whom Johnson is known to have read and admired, George Cheyne may have been reflecting a consensus when, in his *Essay on Regimen* (1740), he maintained that 'it is beyond all Doubt that the Quantity of *solid* Matter in this material *System* is very small, in respect of the interspers'd *Vacuities*'. In a similar manner, Isaac Watts argued, in his *Philosophical Essays on Various Subjects* (1742), that 'whether there be a *Vacuum* or Void Space is now no longer doubted among Philosophers, it having been proved by Sir *Isaac Newton* and others, beyond all Contradiction; and every one agrees to it'. Watts elsewhere asserted that 'the infinite divisibility of matter is so often proved and so universally granted by all modern philosophers, that I need not prove it here'.[5] The sentiments of Cheyne and Watts are much like those of the Dutch chemist, Herman Boerhaave who, like Newton, was alive to the paradoxes inherent in a belief in the infinite divisibility of matter. From the perspective of this belief, Boerhaave argued, the distinction between simple and compound bodies in chemistry becomes difficult to justify; under analysis, it becomes possible to show that simple bodies are susceptible to a potentially infinite proliferation of acts of division: 'the parts which the greatest masters pretend to have resolved compound bodies, are not themselves of a simple nature, but mutable and capable of further division'. An example of this process for Boerhaave can be found in fire; if fire, as Newton had shown, can be divided into seven rays, these rays can in turn be subject to a further division which, in turn, might be liable to further decomposition. Boerhaave was led to conclude that:

> in this vastly simple being, therefore, we see there still remains this manifold variety: What diversity, therefore, have we reason to expect in compounds? In the smallest bodies, we everywhere observe a resemblance to the greater.[6]

The doctrine of the void may thus have the effect of putting in question what is essentially a metaphysical opposition between homogeniety and

heterogeneity, simple and compound bodies: every supposed body may
be shown to be a compound in a process that is arrested only by positing
an invisible point, an absolutely original, primogeneal particle of matter.

These theoretical notions of the vacuum and of the heterogeneity and
porosity of matter may seem remote from Johnson's thought, yet they
deserve our attention if only because they were formulated, as we have
seen, in opposition to the once influential doctrine of the plenum. In its
fully elaborated form, this doctrine was the philosophical foundation of
the principles of plenitude, hierarchy, and continuity that A. O Lovejoy
analysed in his classic *The Great Chain of Being* (1936). Yet because
Lovejoy was apparently unaware of the Leibnizian–Newtonian controversy, he tended to exaggerate the extent to which these principles
were held in common by all eighteenth-century European thinkers. As a
result, he contended that there were only a 'few writers bold enough to
attack the whole assumption of the plenitude of creation. In the second
half of the century, Voltaire and Doctor Johnson and the pioneer
anthropologist Blumenbach were the most notable of these writers'.[7] Yet
one may ask whether Voltaire and Johnson were really that isolated and
whether their ideas were not in fact a development of the Newtonian
argument.

Voltaire's attack on the principle of plenitude is clearly an outgrowth of
his early interest in Newton. This attack was not simply a consequence of
the deepening pessimism that he experienced after the Lisbon earthquake
of 1751. It was an extension of his contention in *The Elements of Sir Isaac
Newton's Philosophy* that 'the Idea of Plenitude is . . . chimerical; that
therefore the whole System, founded on these Imaginations, is no more
than an ingeious Romance, without the Appearance of Truth'.[8] It is only a
short step from this opinion to the claim, advanced in a footnote added to
his *Poem on the Lisbon Disaster* (1755), that 'the chain is not in an absolute
plenum; it has been demonstrated that the celestial bodies perform their
revolutions in an unresisting medium. Every space is not filled. It follows
then, that there is not a progression of bodies from an atom to the most
fixed star. There may of consequence be immense intervals between
beings imbued with sensation, as well as between those that are not'.[9]
This argument is more fully articulated in the chapter on the great chain of
being in the *Philosophical Dictionary* (1752) where Voltaire asks 'how
could you have, in the great empty spaces, a chain that linked everything?
It there is one, it is surely the one that Newton discovered; the one that
makes all the globes of the planetary world gravitate toward one another
in the immense void'.[10]

Johnson's views are more difficult to reconstruct than those of Voltaire since he was more impatient with theories and less likely to couch his philosophical ideas in polemical terms. Yet Johnson's interest in what he described as the vacuity of existence has long been acknowledged; according to Mrs Thrale, 'the vacuity of Life had at some early period of his Life perhaps so struck upon the Mind of Mr Johnson, that it became by repeated Impressions his favourite hypothesis, & the general Tenor of his reasonings commonly ended in that'.[11] Hence it would be surprising, indeed, if the dispute over the plenum and the vacuum that arose in early eighteenth-century England was not among the various literary, philosophical, and religious influences that contribute to this repeated theme in Johnson.

Moreover, the problem of the vacuum — of the distinction between vacuity and plenitude — is of decisive importance for the arguments he advanced in his review of Jenyns's *Inquiry* (1757). A characteristic, if notoriously jejune, contribution to the philosophy of religion in the eighteenth-century, Jenyns's *Inquiry* embraces the principles of plenitude, hierarchy, and continuity in its attempts to provide a rational foundation for religious beliefs. For Jenyns, these principles provide the basis for the claim that evil is an acceptable, even logically necessary, part of the universe. In his review, Johnson not only raises questions about the validity of Jenyns's system itself, but argues, on the grounds of both experience and logic, that it rests on shaky foundations. Thus in his attack on the principle of plenitude, Johnson fastens in the first place on what he regards as the fiction of its accessibility to empirical observation: 'that there is the greatest number possible of every sort of beings . . . with respect to man, we know, if we know anything, not to be true' (*Works*, VI, 52).

The vigour and persuasiveness of this appeal to experience and commonsense are familiar enough to all students of the period. What may be less familiar is the precise bearing of Johnson's critique of the logical premises of Jenyns's argument. Although Johnson never refers to Newton directly, his argument resembles the paradoxical reasoning that Newton employed in treating questions raised by a belief in the porosity and infinite divisibility of matter. By seeking to exclude the vacuum from the universe, Johnson contends, Jenyns only manages to reintroduce it in a way that makes it central and essential:

in the scale, wherever it begins or ends, are infinite vacuities. At whatever distance we suppose the next order of beings to be above man, there is room for an intermediate order of beings between them; and if for one order, then for infinite orders; since every thing that

admits of more or less, and consequently all the parts of that which
admits them, may be infinitely divided.   (*Works*, VI, 53)

There thus emerges an infinite vacuity within the supposed fullness of the
plenum, a self-duplication of division itself. The continuum unfolds into a
scale of being, inseparable from the plenum; and 'there may be room in the
vacuity between any two steps of the scale, or between two points in the
cone of being'. There is a multiplication of vacuities *ad infinitum*,
concludes Johnson for 'infinite power' necessarily entails 'infinite
exertion'.[12]

It is tempting, therefore, to argue that Johnson's notions of plenitude
and vacuity, like those of Voltaire, were firmly set in Newtonian doctrine
and reasoning. The most unusual aspect of his thought on this issue may
well have been his application of Newtonian paradoxes concerning the
vacuity and infinite divisibility not to the internal structure of matter but
to the scale of being. Just as Newton and Boerhaave's studies of matter led
to the notion of heterogeneity, so Johnson's critique of the hierarchy
suggested the idea of chasms, spaces, divisions.

At the same time, however, there is in Johnson's argument a denial of
the assumption that the difference between orders in the chain can be one
of empirical substance. Johnson recognized that the contrast between
infinity and finitude, between perfection and imperfection, necessarily
depends, as Jenyns contended, on the concept of 'limit', but because of the
purely differential, non-substantial nature of the cosmic hierarchy, the
principle of discernibility, Johnson argues, always escapes us. Not even in
the form of a determining 'reason' can we ascertain why a 'being was
suffered to advance thus far and no further' (*Works*, VI, 51–2). Such a
restraint remains ungraspable whether in philosophical, ethical or political
terms.

This may seem like a relatively innocuous argument but, in fact, as
Johnson shows, it challenges the traditional way of thinking about
metaphysics, in which the great chain of being is seen as an a priori given
that is immediately present to the senses. In Jenyns's argument, the chain
of being is only one part of a broader vision in which mind and object,
near and far, part and whole are fused, such that 'the evils suffered on this
globe, may be some inconceivable means contribute to the felicity of the
inhabitants of the remotest planet' (*Works*, VI, 63). Yet as Johnson points
out, Jenyns fails to notice that no human 'conception' can be determined
by 'inconceivable means'. Jenyns's reliance on what is 'inconceivable'
problematizes the relation between part and whole and shows 'speculation' ensnared in a play of language whose gaps and inconsistencies are

every bit as glaring as its logical proofs: 'he has told us of the benefits of evil, which no man feels, and relations between distant parts of the universe, which he cannot himself conceive' (*Works*, VI, 64). Johnson's argument thus combines an interpretive rigour with a perspective that lays bare the extent to which Jenyns's vision fails to conform to ordinary human experience.

Johnson's critique is intended to do more, however, than simply expose Jenyns's philosophical *naïveté*. It is also aimed at producing a drastic change in the way the great chain of being is perceived. Johnson suggests, on the one hand, that the chain is riddled with paradoxes and, on the other, that it is of 'Arabian', origin (*Works*, VI, 49). In this two-pronged assault, Johnson in effect performs for eighteenth-century intellectual history the reverse of what Lovejoy performed for the history of ideas in our own time: he makes the hierarchy seem contradictory as a system and alien and derivative as a current of ideas. Yet this critique, drastic as it certainly is, is not to be taken as wholly undermining the concept. Shorn of its legitimacy in terms of both structure and genesis, the chain of being rather undergoes an intellectual metamorphosis: instead of being a cause, it becomes an effect, a result, a consequence. In place of a system of immutable essences spontaneously and willingly arranging themselves into a vertical order, the hierarchy becomes prey to change, conflict, and discontinuity.

It has sometimes been assumed, on the basis of such arguments, that Johnson's attack on the principle of the hierarchy in his review of Jenyns's *Inquiry* is inconsistent with the stance he adopts elsewhere in his moral and satirical writings. Arieh Sachs has written that 'despite his demonstration that the notion of a universal scale is logically untenable, contradicted by the facts of experience and itself a product of delusive imagination, his satirical exposé of the human condition' depends itself on the idea of man as a middle link in a cosmic hierarchy.[13] A study of the *Review* indicates, however, that Johnson did not so much repudiate the hierarchy as argue that it is incompatible with the principle of continuity. If orders of being are assumed to exist, their existence must be marked by intervals, discontinuities. In contrast to Pope, Johnson never holds that:

> Where, one step broken, the great scale's destroy'd:
> From Nature's chain whatever link you strike,
> Tenth or ten thousandth, breaks the chain alike.[14]

On the contrary, Johnson merely insists that the chain must have 'vacuities, from step to step, through which any order of being may sink into nihility without any inconvenience, so far as we can judge, to the next rank above or below it' (*Works*, VI, 59). There is nothing in this passage which implies that the scale of being will collapse if one of the orders sinks into 'nihility'. Johnson's argument is thus consistent with the traditional scheme that placed man halfway between the angel and the worm. Where it differs from that of Pope, Jenyns, and the continental philosophers who upheld the doctrine of the plenum is in Johnson's insistence that man because of his medial position is separated by a chasm from what is above and below him.

In upholding the principle of the hierarchy, Johnson projected onto the social world the same assumptions that characterized his conception of the cosmic scale of being. Subordination continues to exist in society in spite of the absence of any continuity between the different orders of the hierarchy, but it now becomes subject empirically to historical change, to the play of conflicting forces. One of these forces is 'the lust of dominion and that malevolence which delights in seeing others depressed' (*Works*, VI, 57). The fact that poverty appears as a tragic consequence of this *libido dominendi* means that it should not be given the delusive consolation of a hypostasized entity; still less that it should be viewed as 'irreversible'. The transformation of the scale of being from a meta-physical superstructure into a historical fact reveals that social orders result not only from the peaceful coexistence of rich and poor, as Pope and Jenyns imply, but also from the violent exercise of power and authority. Johnson's method of undermining the divinization of the social hierarchy that he finds in Jenyns's *Inquiry* is to imply that, at crucial historical moments, it may be reversed: 'the maxims of a commercial nation . . . always suppose and promote a rotation of property, and offer every individual a chance of mending his condition by his diligence' (*Works*, VI, 56–7).

The emphasis upon the rotation of property should not be seen as a repudiation of the principle of subordination, then, so much as a dislocation of the conceptual order in which it had been traditionally articulated. What Jenyns proposed as a given, as a constituent element in a speculative system proves to be a product, dependent and derived in ways that strip it of the authority of simple and pure plenitude. Thus transformed, the principle of subordination can no longer be considered the valid basis for a world view but only a special, though extremely important, consequence of an entirely different kind of world view. This holds at least two possible implications for Johnson's thought. First, his attack on the metaphysics of Jenyns's *Inquiry* was neither completely innovative (in Lovejoy's sense of having no precursors), nor was it simply derivative (in the sense of having already existed in the writings of Newton and his popularizers). Second, Johnson's characteristic concerns as a moralist could very well be described as post/traditional or postplenist, in the sense that they are marked by a break with the major traditional form of legitimating truth and authority in the Renaissance and seventeenth century.

The chapters which follow are all concerned with the implications of this rupture with a traditional mode of grounding authority upon Johnson's major texts. What links these texts to Johnson's review of Jenyns's *Inquiry* is, if my argument is plausible, brought to the surface and systematically articulated. It is no accident that Johnson was acutely interested in matters scientific and technical. That is why, as I attempt to show in chapter two, he was able to apply, more or less self-consciously, the lessons of the Newtonian ontological revolution to the realm of moral experience and thought. In chapters three and four, I argue that the issues raised by Johnson's critique of Jenyns's *Inquiry* can be applied to the political worlds of *Irene*, *London*, and *The Vanity of Human Wishes*. The characteristically Johnsonian notion of a universal contest for superiority is here seen as a perfectly logical and consistent alternative to the great chain of being as a legitimating principle. This argument is pursued from a somewhat different angle in my chapter on the *Plan* and *Preface* to *The Dictionary*, where the endless 'maze' of linguistic variation is shown to be the result of the same dislocation of order and degree as the unending struggle for dominion. In *Rasselas*, as I suggest in Chapter 6, Johnson's logical procedures act to undermine distinctions between inside and outside, peace and war, solitude and society. The undecidability that issues forth from these procedures is such as to engender doubts even in such serious discussion as the debate of Imlac and the astronomer over the immortality of the soul. On the other hand, *A Journey to the Western Islands of Scotland* presents a very different case. It combines a rigorous scepticism concerning evidence with a vision of natural scarcity which, I contend in the seventh chapter, is implicitly sustained by Johnson's rejection of the principle of plenitude.

Yet Johnson's philosophical repudiation of the principles of plenitude and continuity does not finally account for the full significance of his major texts. Beyond the philosophical probing is a still deeper experience; beyond the abstract is the personal, the profound feeling of ontological insecurity in the face of a universe in which certain absolutes have been removed.[15] The principle of plenitude sustained the metaphysical world view to which virtually all thinkers of the Renaissance and seventeenth century, regardless of their philosophical persuasion, subscribed. Once that principle was put in question, the writer was faced with the anguishing problem of how to deal not only with specialized, technical questions concerning matter and the void but also with an entire set of cosmic, religious, and social beliefs outside the framework of a generalized theory that could serve to ground those beliefs.

This problem led Johnson beyond the traditional view of the human mind to the formulation of an analogue capable of adjusting traditional values to the questions posed by the Newtonian argument. An account of the human mind commensurate to the principle of plenitude must somehow be transformed into one more closely congruent with the belief that the universe may be a vacuum. But the explicit inclusion of vacuity into our conception of human experience rules out easy attainment of the ethical ideal implicit in the traditional theory — that of a fully active life. For if we once admit vacuity into our conceptions of space and time, it becomes difficult to believe that activity is our natural and spontaneous response to our quest for happiness.

What Johnson desires is not an end to this ideal, not the glorification of a stoic apathy that renders human action unnecessary. Johnson shares Pascal's belief that the universe may be a vacuum, but, like Pascal, he also retains a traditional *horror vacui*.[16] Johnson's fear of idleness and vacancy has been the subject of recent commentary, yet, as I will argue in chapter two, his critique of the principles of plenitude and continuity may have implications that have not yet been explored. If, for example, time is thought of in terms of succession rather than continuity, how might that affect moral values? One good way to approach the implications of Johnson's post-Newtonian critique of the principles of plenitude and continuity is through his application of that critique to long-standing notions of the human mind.

2

Vacuity, Time and Happiness in Johnson's Moral Psychology

Johnson attacks Jenyns's *Inquiry*, observes Joseph Wood Krutch in *Samuel Johnson*, 'not with weapons drawn from the armoury of Christian orthodoxy but logic of the sort Jenyns himself wished to use'.[1] And indeed, Johnson's procedure is to show, with exemplary precision, the extent to which Jenyns's arguments founder on conceptual difficulties of their own making. These arguments not only fail to preserve the metaphysical opposition between plenitude and vacuity but point to the conclusion that the principles of continuity and hierarchy can only be conceived in terms of the chasms they sought to exclude. Yet Johnson proves unwilling to use the contradictions inherent in Jenyns's reasoning as a pretext for constructing a new speculative synthesis capable of explaining the nature of evil more adequately. On the contrary, his strategy is to employ these contradictions in a way that by its nature is opposed to conceptual systematic thought. To register the force of Johnson's critique would thus involve much more than simply confronting Jenyns's ramshackle edifice with yet another instance of post-Newtonian system-building. Indeed, there is good reason to suppose that Johnson would have regarded such an exercise in system-building as merely the restoration of the same hierarchical antitheses of good and evil, truth and falsehood, pleasure and pain, on a new footing. The error of every 'bigot in philosophy', writes Johnson in *Idler*, No. 10, is that he becomes 'entangled in systems', in which these oppositions, far from being kept distinct, are always 'inextricably complicated' (*Idler, Adventurer*, p. 33).[2]

The recurring condemnation of speculative systems in Johnson's writings should not blind us, however, to the fundamental change in emphasis which the doctrine of the vacuum wrought in his moral psychology. Unlike many thinkers of the era, Johnson never explicitly claimed to apply the methods and assumptions of Newton to areas outside the domain of natural philosophy. Indeed, this may be the reason why he refused even to mention Newton explicitly, when the occasion arose, in *Rambler*, No. 8. To Johnson, such a contention would

11

undoubtedly have seemed as pretentious as Jenyns's claim to have formulated a new explanation of the nature of evil. None the less, Johnson was never reluctant to put forth his own convictions in terms of categories drawn from the New Science. As W. K. Wimsatt, Jr has observed, Johnson freely appropriated the terminology of seventeenth and early eighteenth-century science to suit his own ends, applying it 'to psychology in a metaphoric manner'.[3] Part of that process may involve Johnson's application of the terminology of his critique of the principles of plenitude and hierarchy to the moral concerns of his essays in *The Rambler, Idler,* and *Adventurer* essays.

To understand how Johnson might have adapted the vocabulary of his critique of the principle of plenitude to psychology, it is helpful to compare his conception of the human mind to that of René Descartes. Even if Descartes' own metaphysical system was rejected by later philosophers, it was based on the same doctrine of the plenum as the systems of Leibniz, Pope, and Jenyns. Descartes' particular version of the plenum, moreover, extended to mind as well as matter. For Descartes, every conscious state is a 'thought'. A 'thing which thinks' is a 'thing which doubts, understands, affirms, denies, wills, refuses, which also imagines and feels'.[4] In this extremely broad understanding of thinking, Descartes implies that the mind is always 'full' in two senses; it is full because it is always active whether it is waking or sleeping, and it is full because its thoughts are always immediately present to itself. Of course, Descartes' notion of a fully conscious and active mind is an inseparable part of his argument for the existence of the *cogito*. But his application of the plenum to human psychology was not novel; similar applications can be found in the works of seventeenth-century English writers. John Donne, for example, had argued that:

> *Solitude* is a torment which is not threatened in *hell* it selfe. Meere *vacuitie*, the first *Agent*, God, the first *Instrument* of God, *Nature*, will not admit; Nothing can be utterly *emptie*, but so neere a degree towards *Vacuitie*, as *Solitude*, to be but one, they love not.[5]

No doubt such sentiments were basic to Donne's and all notions of the human mind based on the doctrine of the plenum. But they are not necessarily to be found in theories of the mind that depend on a belief in the void. By way of contrast to Descartes and Donne, George Cheyne, one of the era's most ardent Newtonians, clearly assumes that there are periods when the mind may be vacant and inactive. In his *Philosophical Principles of Religion* (1715), Cheyne envisages a condition in which 'the *raional Soul* is but weak, faint, and languid, and almost void of all *Ideas* and

Images, these being not only separable, but at least to be *actually* separated'.[6] This condition arises because the mind exists in a constant state of vulnerability as it searches for objects able to satisfy its infinite desires. Inactivity in this theory does not simply result from the sin of sloth, which is the way the plenist perceives it. Rather, inactivity is the inescapable consequence of the instantiation in the psychic world of the disparity between infinite void and finite matter that Newton imagined to be the essential condition of the natural world. In Cheyne's theodicy, this disparity can only be overcome when the human mind discovers in the divine mind a plenitude commensurate with its aspirations:

Desire . . . is as the *Infinity* of *Space* to the divine Plenitude: which infinite Space, nothing created can adequately fill, but the divine *Plenitude*. And in this view, the *infinite Capacity* of the *Desire* may be considered as a *boundless Void*, made to receive some *fleeting*, limited Parts or Systems of Matter . . . but can be perfectly filled, or adequately replenished, but by the *supreme Infinite*: Who is perfect with, and replenishes every Point of the great and *Universal Void* of Nature.[7]

Of course, there are plenty of precedents for the idea that man's desire for happiness can only be satisfied in God. The notion was a commonplace in Renaissance and seventeenth-century thought. But in Cheyne's version of this commonplace, the distance between human desire and spiritual fulfilment is radicalized by its incorporation into the framework of an opposition between an infinite inner space and a divine plenitude. That Johnson envisioned a similar application of the Newtonian doctrine of interstitial space to the human mind is made clear in *Rambler*, No. 8. Where he differs from Cheyne is in the extent to which he brings out the profoundly disturbing implications of this conception of the human mind for temporal existence. From the very first he describes that inner space in negative terms, characterizing in *Rambler*, No. 8 and elsewhere as a void and a lack. In much the same way that if 'all matter were compressed to perfect solidity, it might be contained in cube of a few feet', so Johnson argues, if 'all the employment of life were crowded into the time which it really occupied, perhaps a few weeks, days or hours, would be sufficient for its accomplishment, so far as the mind was engaged in the performance' (*Rambler*, I, 41). Johnson also proceeds to extend this estimate of 'the time . . . the employments of life . . . really occupied', to the realm of reflection, claiming that if even:

the most active and industrious of mankind was able, at the close of life, to recollect distinctly his past moments, and distribute them, in a regular account, according to the manner in which they have been spent, it is scarcely to be imagined how few would be marked out to the mind, by any permanent or visible effects, how small a proportion his real action would bear to his seeming possibilities of action, how many chasms he would find of wide and continued vacuity, and how many interstitial spaces, unfilled, even in the most tumultuous hurries of business, and the most eager vehemence of persuit. (*Rambler*, I, 41)

The troublesome implications of this argument are evidenced in the discovery not only of chasms and interstitial spaces in even the most productive existence but in the disclosure of an essential quiescence at the very heart of human action. The busiest moments of an industrious person's life, no less than his empty hours, are a modality not of plenitude but of vacuity. As spatial continuity implies vacuities, moreover, so temporal activity necessarily entails intervals of inactivity. In these intervals, the mind possesses the attribute of being able to exist without any motion or power whatever, its activity dwindling to the point where it is barely conscious of its own existence. Incapable of receiving the imprint of any 'external impulse', or of reflecting upon past or future, the mind sinks at this point 'into a state approaching that of brute matter, in which' the individual 'shall retain consciousness of his own existence only by an otiose langour and drowsy discontent' (*Idler*, *Adventurer*, p. 31). Though langour and discontent are evidence that the mind has retained a certain minimal awareness, it can no longer presume to rest, in the Cartesian sense, on a knowledge of its own existence in the present act of thinking.

But vacuity is not only presented as a constituent element of mental experience. It also serves as a boundary condition or constraint on all human behaviour, saddling man, for example, with the imperative not only of filling up the empty spaces of solitude but of minimizing the vacuities that inhabit even the busiest moments. This is one of the areas of course where Johnson's outlook is intensely personal.[8] He does not simply take up contemporary issues and then attempt, like Hume, to resolve them in an ironically detached or disinterested manner. These issues arise out of his own life in the sense that they give shape and definition to his own deepest fears and terrors. In this sense, Johnson can be said not merely to have generalized his own experience of mental blankness and stagnation in his *Rambler*, *Idler*, and *Adventurer* essays but to have 'lived' the dilemmas of his age. A preoccupation with vacuity, for example, brings Johnson not only to conventional considerations of the deleterious effects of idleness on man, but to the broader question of what constitutes an active life insofar as that phrase points to a contrast between a state of movement and a state of quiescence, between activity and inactivity, action and inaction. What is imponderable about this question, moreover, is the possibility that what is commonly called 'business' or 'employment' does not always signify true 'labour'. Thus a part of Johnson's anxiety rests on the belief that a full life is an aim that can seemingly be translated into practice by a 'state of action' that is in reality devoid of plenitude – a fact that renders the self's reflection on its own activities disconcertingly uncertain and ambiguous.

'Real employment' is one of a series of phrases used in the *Idler* essays to indicate the productive structure of authentic activity. An examination of the various types of activity, and even more importantly, of the values that Johnson attributes to them, would show that he places a high premium on actions involving the use of tools and directed toward the outside world. To observe, to take, to seize, to shape, to transform: these are useful activities that can fill up the intervals of even the most humdrum existence. In Johnson's view, their value lies not merely in the transformation of material objects but in the exercise of 'close thought and just ratiocination' (*Idler, Adventurer,* p. 97). Yet because manual activity is a disjunctive event, forever recurring and never coming to an end, it is subject to the self-deception involved in the idleness that simulates activity. In *Idler,* No. 31, Johnson describes this idleness in terms of deferral: 'Nothing is to be expected from the workman whose tools are forever to be sought'; and dispersion: Mr Sober 'has attempted at other times the crafts of the shoemaker, tinman, plumber and potter' (*Idler, Adventurer,* pp. 96, 97). In these futile efforts, Johnson may be describing his own traits: they depict a self abandoning its mind to the illusory plenitude of its own efforts. There is no longer any actual contact with reality; its norm has become not practical utility but 'daily amusement'.

The crucial point here is not merely that there is a close link between Johnson's own life and his unmasking of the various forms of deception idleness can take. It is that his argument is defined in terms of assumptions that place the active life beyond the realm of human achievement. Indeed, in the paradox of an ethics of activity whose protagonists are mainly conscious of their inactivity lies one of the most striking features of Johnson's moral writings. How can one be expected to aspire to a 'full' life when plenitude necessarily inhabits and is inhabited by vacuity? At the risk of repeating what may seem familiar to some readers, the answer lies not in the emptiness of the inner space (or time) to be filled, but in the incessant inclination that seeks to fill that emptiness.

What makes this inclination another source of disquietude, however, is that it is the embodiment for Johnson, not of a certain psychic state to which individuals 'naturally' tend, but of a compulsion which they are forced to obey: 'those who have already all that they can enjoy, must enlarge their desires. He that has built for use, till use is supplied, must begin to build for vanity, and extend his plan to the utmost power of human performance, that he may not be soon reduced to form another wish' (*Rasselas*, 32, 113–14). Johnson refers to this inner compulsion as the 'hunger of imagination that preys incessantly on life'.[9] From it there follows the discovery, central to Johnson's moral outlook, that the final goal of this restless search is always deferred. Once again, plenitude and vacuity prove to be indivisibly bound. Anticipated delight is actual satiety. Given the fear which this discovery engenders, it becomes easy to understand why the imagination is so fragile: the possibility that its quest might prove empty fosters doubt, encourages the pursuit of 'ease', and invites the mind to dwindle into a 'state of unruffled stupidity' (*Idler, Adventurer*, p. 96).

The subjective correlative of this dilemma shows up most clearly in the *Prayers and Meditations*. Here personal experience matters, since it is Johnson's account of mental states that were barely states which is the puzzle – and perhaps the key to the difficulty that contemporary readers encountered in reading Johnson's record of his personal trials. For, as W. B. C. Watkins noted, 'the anguish of soul therein revealed seemed to his contemporaries out of all proportion to the degree of sin which caused it'.[10] Yet there can be no mistaking the depth of this anguish, as for example, when Johnson writes 'I am now to review the last year and find little but dismal vacuity, neither business nor pleasure' (*Diaries, Prayers*, p. 294). Not being in the world, this 'I' could hardly have existed except in reflection. The moral dilemma was thus also a psychological dilemma, one that becomes even more striking since it barely touches on the issue of sinfulness in any conventional sense: 'a kind of strange oblivion has overspread me, so that I know not what has become of the last year, and perceive that incidents and intelligence pass over me without leaving any impression' (*Diaries, Prayers*, pp. 77–8). In the spontaneous dovetailing of this apparent immobility with Johnson's endlessly restless desire for felicity lies the unique difficulty to which many of the prayers and meditations are devoted. It is typical of Johnson to characterize his wish for an alleviation of this difficulty as a desire for consciousness. Because consciousness is ultimately consciousness of one's thoughts, feelings, and actions, a prayer for mercy becomes a prayer for self-identity.

*   *   *

The disparity between man's incessantly active quest for happiness and his capacity for inactivity is clearly shown not only in Johnson's anxieties about the vacuities of his own life but in everything he wrote about human experience. Certainly this assumes a different point of view than would be found in moral theories based on the plenum. Rather than commencing with *a prioris* about the inexhaustible variety and abundance of objects available to satisfy man's desires, Johnson begins with a conception of life as a vast emptiness needing to be filled up. W. J. Bate has drawn attention to the importance of the verb 'fill' in Johnson's moral essays.[11] The application of the imperative implicit in this verb to the moral realm produces a characteristically Johnsonian dilemma: the fact, as he sees it, that 'so few of the hours of life are filled up with objects adequate to the mind of man' (*Rambler*, I, 221).

But this dilemma leads Johnson in his essays, not to a consideration of scarcity and its implications, as we might expect, but to the quite different issue of the nature of temporal experience. Even if we want to assume that there are enough objects to fill up 'the hours of life', the hours themselves prove, in Johnson's argument, to be complex constructions. Indeed, temporal duration becomes a ground for plenitude only insofar as it is marked by the same vacuities that divide the different orders in the scale of being. Thus, just as there are two rival conceptions of space in Johnson's critique of Jenyns's *Inquiry*, so there are two alternative ideals of time in his *Rambler* and *Idler* essays: continuity and succession. The idea of time as continuity is consonant, not with a 'full' life, as one might suppose, but with a state of inertia. Variously described in Johnson's essays as a 'stream', 'a current', or an 'even and unvaried tenor of life' that 'glides unseen and unfelt', this idea of time is subject to no exteriority, no alterity, not even that of a beginning or an ending. Hence it is not surprising that, as Johnson puts it in *Idler*, No. 103, a person who 'lives today as he lived yesterday, and expects that, as the present day is, such will be the morrow, easily conceives time as running in a circle and returning to itself' (*Idler, Adventurer*, p. 315). Such a person will easily come to believe that the finite movement of time obliterates all differences, that it reproduces itself in an endless repetition of the same.

To avoid the illusion that time 'is running in a circle', one must apprehend it as a succession of discrete moments, of intervals. This idea of time bears a striking resemblance to what David Hume argued in *A Treatise of Human Nature* (1742):

as time is composed of parts that are coexistent, an unchangeable object, since it produces none but coexistent impressions, produces none that can give us the idea of time; and, consequently, that idea must be derived from a succession of changeable objects and time in first appearance can never be severed from such a succession.[12]

Maintaining, like Hume, that 'succession is not perceived but by variation' (*Idler, Adventurer*, p. 315), Johnson insists that

> if the wheel of life, which rolls thus silently along, passed on through undistinguishable uniformity, we should never mark its approaches to the end of the course. If one hour were like another; if the passage of the sun did not shew that the day is wasting; if the change of seasons did not impress upon us the flight of the year, quantities of duration equal to days and years would glide unobserved. If the parts of time were not variously coloured, we should never discern their departure or succession, but should live thoughtless of the past, and careless of the future, without will, and perhaps without power to compute the periods of life or to compare the time which is already lost with that which may probably remain.   (*Idler, Adventurer*, pp. 135–6)

Consciousness, from this perspective, is related to pauses, interruptions, 'points of time where one action ends and another begins' (*Idler, Adventurer*, p. 315). Since these are points of time which can prompt us to reflect on the last point of time, the final course of action, our impulse is to suppress them, to abandon ourselves to the false plenitude of our imaginings. When we give into this temptation, 'we suffer phantoms to rise up before us, and amuse ourselves with the dance of airy images, which after a time we dismiss for ever, and know not how we have been busied'. For those who habitually indulge in such fancies, past and future dissolve into a single, continuous present. Johnson describes their fancies as a dream of pure plenitude, a desire 'which sometimes puts sceptres in their hands or mitres on their heads, shifts the scene of pleasure with endless variety, bids all the forms of beauty sparkle before them, and gluts them with every change of visionary luxury' (*Idler, Adventurer*, p. 101). When the mind abandons itself to this 'circle of happiness' it allows no interruptions, no intervals, erasing time and its differences into total self-presence. In such visionary moments, the mind can easily transform itself into the prime mover, the lord who, controlling himself, controls time and space:

It is easy in these semi-slumbers to collect all the possibilities of happiness, to alter the course of the sun, to bring back the past, and anticipate the future, to unite all the beauties of all seasons, and all the blessings of all climates, to receive and bestow felicity, and forget that misery is the lot of man. (*Idler, Adventurer*, p. 101)

This transformation of self-presence into omnipresence, of the dreaming subject into the supreme being is a characteristic feature of what Johnson defines as 'madness'. It is most apparent of course in the astronomer of *Rasselas* who, seemingly against his own will, comes to believe that he has gained dominion over the weather. For Johnson, the astronomer's illusory happiness consists in the pleasureable consumption of an absolutely enclosed plenitude. Lacking anything 'external that can divert him', the astronomer 'must find pleasure in his own thoughts, and must conceive himself what he is not, for who is satisified with what he is? He then expatiates in boundless futurity, and culls from all imaginable conditions that which for the present moment he should most desire, amuses his desires with impossible enjoyments and confers upon his pride unattainable dominion' (*Rasselas*, 42, 140). But it turns out, needless to say, that the astronomer's plenitude is only another form of the vacancy that Johnson describes so powerfully in the *Prayers and Meditations*. Indeed, the present moment can serve as the ground of the astronomer's happiness only insofar as it is not a pure and immediate presence. For the mind to avoid madness, it must insure that its plenitude is marked by vacuity, its identity by difference, its interiority by exteriority.

The capacity of the mind to generate the illusion of plenitude problematizes the notion of the present moment for Johnson. When time is apprehended as continuity, the present moment comes to appear as an eternal moment, an 'endless duration' in which present and 'boundless futurity' merge into one. Indeed, Johnson defines eternity in *The Dictionary* as 'duration, without beginning or end', illustrating his definition with a quotation from Richard Crashaw that describes this duration in terms of 'circular joys/Dancing an endless round'. When time is apprehended as succession, on the other hand, the present moment comes to divide itself into before, now, and after; past, present, and future. But once time is perceived as succession, the present becomes, not the moment of presence, but that which is no longer or not yet. Indeed, Johnson argues that 'almost all that we can be said to enjoy is past or future: the present is in perpetual motion, leaves us as soon as it arrives, ceases to be present before its presence is well perceived, and is only known to have existed by the effects which it leaves behind' (*Rambler*, I, 223–4). Hence the 'present' cannot be said to participate in 'presence'; it can no longer be apprehended as a simple, indecomposable absolute, but rather by the 'effects' which it produces.

It is thus this argument, not reflections on scarcity, that provides the philosophical framework for the contention that 'so few of the hours of life are filled up with objects adequate to the mind of man'. Here the issue is still the hierarchical opposition plenitude/vacuity. Johnson's reasoning involves the implication that as a fleeting instant, the present moment never exists in an isolated moment of awareness. Value thus lies not in what is given to the mind in the here and now but in what is added or what supplements the deficiency within the given:

> We are forced to have recourse every moment to the past and future for supplemental satisfactions and relieve the vacuities of our being, by recollection of former passages, or anticipation of events to come.   (*Rambler*, I, 221).

Past and future can thus be compensatory to the present, only because the present is already marked by the qualities which are conventionally predicated of the past and future; absence and vacuity. But it soon becomes clear that in Johnson's moral psychology past and future are not equally weighted. The past can never be retrieved in the form of a revivified present, because the past is remembered only to the extent that it is marked as past, as lost and irretrievable: 'every revived idea reminds us of a time when something was enjoyed that is now lost' (*Idler, Adventurer*, p. 139). Moments can be recollected, that is to say, only if they are remembered, not as something immediately given, but as the products of a contrast between then and now. Characteristically, Johnson seeks to support this philosophical argument by an appeal to personal experience:

> so full is the world of calamity, that every source of pleasure is polluted, and every retirement of tranquillity disturbed. When time has supplied us with events sufficient to employ our thoughts, it has mingled them with so many disasters, that we shrink from their remembrance, dread their intrusion upon our minds, and fly from them as from enemies that pursue us with torture.   (*Rambler*, III, 291–2).

Thus, even the present moment is marked by the same link between joy and sorrow, pleasure and pain, that characterizes a past moment as past. This linkage means that pleasure and joy are only constituted in relation to their opposites; neither pleasure nor joy is simply present or absent, but is invariably entangled with pain or sorrow. The consequence for ethics is the virtual collapse of all those oppositions that Jenyns set up in order to explain the origin and nature of evil.

* * *

The breakdown of these oppositions requires nothing less than the recasting of our traditional assumptions about life, making it much more difficult for us to suppose that any part of temporal experience can serve as the foundation of pure felicity. This supposition leads Johnson to contest the view that our 'hopes' for the future can remedy the deficiencies of past and present. Thus even if it is true that 'every period is obliged to borrow its happiness from the time to come' (*Rambler*, III, 293–4), this borrowed happiness is also intrinsically ambivalent. While avoiding the fallacy of a vacuous inertia, it none the less announces itself as little more than a strategy or delay, deferral, and substitution. It is clear that Johnson regards the imperative behind this strategy of substitution as universal.

> every man is sufficiently discontented with some circumstances of his present state, to suffer his imagination to range more or less in quest of future happiness, and to fix upon some point of time, in which, by a removal of the inconvenience which now perplexes him, or acquisition of the advantage which he at present wants, he shall find the condition of his life very much improved.   (*Rambler*, I, 25)

Johnson's empiricism sometimes leads him, as it does in this passage, to qualify a sentiment by quantitative determinations – e.g. 'some', 'more or less', 'very much'– which suggest the possibility of individual variation. But this possibility seems very remote in Johnson's account of our universal 'quest of future happiness', for what the future discloses is the impossibility of ever possessing the anticipated object: 'when this time, which is too often expected with great impatience, at last arrives, it generally comes without the blessings for which it was desired; but we solace ourselves with some new prospect, and press forward again with equal eagerness' (*Rambler*, I, 25). What the future actually produces therefore is an endless chain of substitutions which, at best, generate only a simulacrum of the pleasure they defer:

of riches, as of everything else, the hope is more than the enjoyment; while we consider them as the means to be used, at some future time, for the attainment of felicity, we press on our pursuit ardently and vigorously, and that ardour secures us from weariness with ourselves; but no sooner do we sit down to enjoy our acquisitions, that we find them insufficient to fill up the vacuities of life.   (*Idler, Adventurer*, p. 228)[13]

The notions of hope as a supplement and a deferral are commonplaces which doubtless have their roots deep in traditional Christian homiletic. Indeed, Chester Chapin has found in Johnson's preoccupation with such notions proof that he, like Pascal, was engaged in his essays in formulating a 'psychological' argument for the existence of a Christian god. Pointing 'to the unhappiness of human life as evidence of man's transcendent destiny', this argument, Chapin holds, presupposes 'the reality of the Fall and original sin'.[14] Yet while there is no denying that Johnson conceives of 'man's transcendent destiny' as a recuperation of the plenitude missing in this life, it is less certain that his view of life actually assumes the doctrine of original sin. For his elaboration in the *Rambler* and *Idler* essays of the themes of supplementarity and substitution – like his critique of the principle of plenitude in his review of Jenyns's *Inquiry* – seems to depend more on reason than revelation. Typically, the notion of the indefinite deferral of desire appears as an ultimate subversion of the attempt to categorize the general system of temporal existence into one or the other of two separate and mutually exclusive domains.

Moreover, this disruption of conventional categories can itself become the means by which pleasure is introduced into the workings of the system. Happiness and misery, far from being opposed as they are in the theology of Jenyns, are in some sense seen as inextricably intertwined in actual life. Under the inexorable pressure of advancing age, it is true, the element of pleasure can disappear, leaving 'religious hopes' as the only counterpoise to the miseries that Johnson describes so graphically in *Ramblers*, Nos. 69 and 203. But in the earlier ages of life, Johnson's argument turns on the element of uncertainty, which inhibits any clear-cut distinction between pain and pleasure, sorrow and happiness. From this point of view, Johnson can thus argue that even secular hope is valuable, 'tho hope should always be deluded, for hope itself is happiness, and its frustrations, however frequent, are yet less dreadful than its extinction' (*Idler, Adventurer*, p. 182). It is perhaps because of the ambivalence inherent in the compounding of these two radically contrasting notions that Johnson defines hope, in *Adventurer*, No. 69, as a 'cordial' that can act as a cure and as a soporific:

> so scanty is our present allowance of happiness, that in many situations
> life could scarcely be supported, if hopes were not allowed to relieve
> the present hour by pleasures borrowed from futurity; and reanimate
> the languor of dejection to new efforts, by pointing to distant regions
> of felicity, which yet no resolution or perseverance shall ever
> reach.   (*Idler, Adventurer*, p. 394)

Hope is thus presented as an antidote, as a remedy to the 'languor of
dejection'; yet this antidote is also a narcotic:

> But these, like all other cordials, though they may invigorate in a small
> quantity, intoxicate in a greater; these pleasures, like the rest, are lawful
> only in certain circumstances, and to certain degrees; they may be
> useful in a due subserviency to nobler purposes, but become dangerous
> and destructive, when once they gain the ascendant in the heart; to
> sooth the mind to tranquillity by hope even when that hope is likely to
> deceive us, may sometimes be useful; but to lull our faculties in a
> lethargy, is poor and despicable.   (*Idler, Adventurer*, p. 228)

In referring to 'hope' as a 'cordial', Johnson is thus not making a simple
value judgment. In *The Dictionary*, he defines a cordial as 'a medicine that
increases the force of the heart, or quickens circulation', yet he also cites a
quotation from *Arbuthnot on Ailments* to the effect that 'a cordial, properly
speaking, is not always what increaseth the force of the heart; for, by
increasing that, the animal may be weakened, as in inflammatory
diseases'. In *Adventurer*, No. 69, the distinction between proper and
excessive dosages of a cordial is assimilated to a conventional distinction
between restraint and excess, reason and imagination. Yet these polarities
may easily become inverted, since the cordial has no proper and
determinate character as a medicine and indeed is composed of the very
same substance as the disease itself.

It might be argued that the ambivalence inherent in Johnson's view of
hope as a cordial is suspended whenever he refers to religion ('the great
task of him, who conducts his life by religion, is to make the future
predominate over the present' (*Rambler*, I, 37–8). Yet not even this 'hope'
may be immune to the 'snares . . . by which imagination is intangled'
(*Rambler*, I, 45). These snares are found, however, not in the satiety that
accompanies enjoyment, but in the danger that attends anticipation. Thus
we cannot discuss our hope for a future state, Johnson implies, without

taking note of the way our minds ceaselessly oscillate between 'presumption' and 'despondency', 'heady confidence' and 'heartless pusillanimity' (*Rambler*, I, 137). Here, as throughout the passages on temporal hope, affirmation engenders – however tenously or provisionally – the possibility of negation. And vice versa; the ambivalence of the cordial and of hope is inescapable.

These same oppositions and reversals of value also apply to the present moment. For past, present, and future, though distinguishable, are none the less only aspects of the present moment. As a result, the person who seeks to anticipate the future is one who must divide himself, must cease to exist fully in the present. Becoming absorbed in its contemplation of a prospective event, the mind risks the danger of neglecting the here and now. As Imlac warns Rasselas and Nekaya, 'it seems to me . . . that while you are making the choice of life, you forget to live' '(*Rasselas*, 30, 108). This does not mean that we should abandon hope, for hope, as we have seen, is itself a happiness. But through an unexpected inversion this happiness is only prospective because, as Johnson argues in *Idler*, No. 58, 'pleasure is seldom found where it is sought. Our brightest blazes of gladness are commonly kindled by unexpected sparks' (*Idler, Adventurer*, p. 180). In a word, we seem to span two worlds, one of the present, which is only pleasurable when it is surprising in the suddenness of its appearance, and one of the future, to which the delights of anticipation are more germane. Moreover, our situation is rendered doubly difficult in that we are unable to rely on the plenitude of the present, nor consign ourselves wholly to the happiness of an anticipated future. Indeed, it may be that the most striking feature of Johnson's moral psychology lies in the fact that happiness and pleasure are always fugitive – they disappear in the very act of appearing, and appear only in the act of disappearing.

*   *   *

Taken by itself, this denial of a privileged ground of pleasure or happiness in temporal experience suggests that pessimism might very well be the appropriate term to characterize Johnson's attitude toward life. And indeed, Johnson's ethical outlook is pessimistic in the sense that an adding up of the quantities of pleasure and pain on the balance sheet of existence would give, in Johnson's reckoning, an excess of pain. But it should be emphasized that pessimism does not mean only this for Johnson. It means, in addition, that pleasure itself is a compound, never existing in pure and simple plenitude. In the present context, this does not involve, as has already been remarked, asserting the absolute primacy of vacuity over plenitude, of misery over happiness. Nor does it involve denying pleasure to common human activities. But it maintains that these activities in general depend on sources whose susceptibility to error always puts their value in question.

In the human psyche, Johnson distinguishes two main sources of pleasure, the eye and ear, the second being much more prone to delusion than the first. In *Rambler*, No. 5, Johnson describes how the mind can overcome the satiety of the present moment without having recourse to the future by transforming itself into an 'eye'. From Johnson's standpoint, this involves 'flying from oneself' and becoming 'open to every new idea'. Everything comes to be placed before this eye which, shifting its gaze outward, turns from an 'idler' into a 'rambler'. Yet 'it is to no purpose' if the 'subject alters his position' while 'his attention remains fixed to one point' (*Rambler*, I, 28). The astronomer in *Rasselas* is one of those whose 'thoughts have been long fixed upon a single point' and have thus acquired a 'particular' cast (*Rasselas*, 40, 134). This concentration of focus, indeed, can constitute itself only through the suppression of all novelty, all heterogeneity. To see, for Johnson, thus means not only to turn outward but to take cognizance as well of the ineradicable difference between monotonous repetition and the kind of repetition founded not on identity but on the reiterated renewal of the experience of variety.

Yet the variety which the attentive eye perceives in the external world provides a striking contrast, in Johnson's essays, to the uniformity which the ear, the organ of illusion, can sometimes experience sounds. Unlike the eye, the ear can listen to the mind, can hear its voice speaking to itself, without having to make a detour through the exteriority of the world. Here is a typical passage on the voice of idleness:

> Vanity, thus confirmed in her dominion, readily listens to the voice of idleness, and sooths the slumber of life with continual dreams of excellence and greatness. A man elated by confidence of his natural vigor of fancy and sagacity of conjecture, soon concludes that he already possesses whatever toil and enquiry can confer. He then listens with eagerness to the wild objections which folly has raised against the common means of improvement: talks of the dark chaos of indigested knowledge; describes the mischievous effects of heterogeneous sciences fermenting in the mind; relates the blunders of lettered ignorance; expatiates on the heroick merit of those who deviate from prescription, or shake off authority; and gives vent to the inflations of his heart by declaring that he owes nothing to pedants and universities. (*Rambler*, III, 56).

Between what vanity hears and what idleness speaks, neither interruption, nor objection appears to interpose itself. But such a closed circuit inevitably becomes a source of error because there is no actual accession of knowledge or contact with the outside. Rather it fosters the illusion, through the persuasive power of its voice, that the outside is inside, indeed that its audience encompasses the cosmos. Under the spell of such an illusion, the astronomer confesses to Imlac: 'the sun has listened to my dictates, and passed from tropick to tropick by my direction; the clouds, at my call, have poured their waters, and the Nile has overflowed at my command' (*Rasselas*, 41, 136).[15]

It is in this way that the sonority of the voice contributes to the river by which the mind allows itself to be carried around and around in the circle of temporal wish-fulfilment. The role of the voice in establishing this circle is further confirmed by its link to the lust of dominion. The interiorization of the voice strengthens individual resolve, as the autonomous circuit of hearing-oneself-speak excludes any possible intrusions from without. But the interior ear, it turns out, can never achieve genuine plenitude because the 'ardour of enterprize' requires for its satisfaction the applause of those very voices whose objections it has stilled or ignored. The attempt of the solitary voice to exclude the voices of others depends, we find, upon its having already interiorized these voices within its field of consciousness. Thus even when it dwells in lonely splendour, the voice never quite attains the solipsistic isolation of hearing-itself-speak but is always engaged in a debate or confrontation, whether real or imaginary, with the world.

In his articulation of this limitation in the notion of voice as presence, Johnson points to the crucial blind spot inherent in the 'lust of dominion': its dependence on others for the validation of quests that are supposed to transcend dependency. In chapters three and four, I will try to show the different ways this contradiction undermines the deluded overreaching of empirical limits in *Irene, London,* and *The Vanity of Human Wishes.* And the interiorization of the outside lies at the root, I shall argue in Chapter 6, of the issue of solitude and society, as waged between Imlac, the hermit, and astronomer in *Rasselas.* Suspended undecidably between two poles, two rival modalities in the choice of life, Johnson's Oriental tale can only repeat their difference through constantly reduplicating the narrative of its own philosophical quest.

3

Desire, Emulation and the Dialectic of Domination and Servitude in *Irene*

Desire is the term which Johnson employs to describe the motive force by which the mind is driven to break out of the circle of idleness and illusion and engage in a confrontation with the world. Whereas the perceiving consciousness of the 'spectator' or 'rambler' seeks merely to study and comprehend the world, the desiring mind of the 'adventurer' seeks to appropriate it, to make it its own. Johnson's most detailed discussion of desire occurs in *Rambler*, No. 49. Paul Alkon observes that *Rambler*, No. 49, contains 'a very full treatment of "mental anatomy"', making an important distinction between the 'natural passions that are universal constants in human nature' and the 'artificial passions that do not necessarily appear or play the same part in everybody'.[1] Johnson's distinction between natural and artificial passions comprise two out of what might be described as four stages in the growth of the human mind. These stages encompass the movement of the mind from the natural appetites of the infant in the first stage to the universal struggle for emulation that is characteristic of the adult in the final stage.

This conception of desire seems quite close to what George Cheyne described in the second edition of his *Philosophical Principles of Religion*. Cheyne, as we have seen, extended the Newtonian conception of infinite void space to the human mind, contending that desire is a want or a lack that can never achieve temporal satisfaction. In a similar manner, Johnson locates desire in *Rambler*, No. 49, on the horizon of what he calls a 'state of inactivity', implying that this state can never be fully encompassed by any finite object or natural need. In the political realm, this conception of desire as a never-completed movement toward completion is seen as both producing and undermining the structure of subordination that is present for Johnson in all societies. Rather than contesting the principle of subordination – as desired by some of Johnson's more liberal contemporaries, with whom he was wholly out of sympathy – Johnson strives in his major poems to articulate the different power relations that lie behind all systems of subordination. Within such systems, desire is manifested for Johnson in a perpetual struggle for superiority that can only be crystallized when the members of a society sort themselves out into an unstable yet necessarily repressive hierarchy of strong and weak, eminent and obscure.

Johnson's conception of civil society was worked out in relation to his notions of desire and inactivity. As an emptiness that can never be filled, 'inactivity' inevitably becomes a source of 'uneasiness', a motive for action. The close relation between desire and 'uneasiness' – a linkage that serves to support Johnson's contention that good and evil, pleasure and pain, are inextricably mingled in human life – provides the basis for the articulation into pairs of the six natural passions of the second stage of Johnson's anatomy of desire in *Rambler*, No. 49: 'hope and fear, love and hatred, desire and aversion'. Arising from the 'power of comparison and reflexion', these passions gradually extend their 'range', until the mind can apprehend not only immediate pain but, in a distinction that resembles the Newtonian conception of force, 'fear at a distance'. For Johnson, the mind is already dominated in this phase by a 'terror' that can be transformed into 'caution' and thus has already acquired, whether through 'reason or experience', the calculation that will enable it to 'endure many things in themselves toilsome and unpleasing' in order to 'procure some possible good, or avert some evil greater that itself' (*Rambler*, I, 264). The predominance of fear over desire in this stage is an indication that its primary object is only the unheroic desire to survive. Yet even this limited biological goal proves impossible for man, for he learns that 'the wants of nature are soon supplied' and that 'something more is necessary to relieve the long intervals of inactivity, and to give those faculties, which cannot lie wholly quiescent, some particular direction' (*Ramlber*, I, 264). This deliberate reintroduction of inactivity gives rise to a new phase of existence in which 'artificial' passions, though wholly extrinsic, become a necessary supplement to the deficiency of the natural passions.

But the explicit introduction of 'artificial' passions, with their accompanying logic of supplementarity, also works a fundamental change in the nature of desire itself. With the rejection of biological existence as a fixed datum, it is no longer possible to assume that the world exists for the mind only as a means. Now that biological existence has been superseded, the principle of utility is replaced by a new logic of conventionalism: 'we persuade ourselves to set a value upon things which are of no use, but because we have agreed to value them '(*Rambler*, I, 264). The principle of utility now ceases to be the *a priori* that some critics have found in Johnson's general theory of human nature and becomes instead a set of instructions to be carried out on behalf of impulses which, in themselves, can by no means be regarded as necessarily advantageous or beneficial: 'from having wishes only in consequence of our wants, we begin to feel wants in consequence of our wishes' (*Rambler*, I, 264).[2]

This introduction of artificial passions in the very heart of mental anatomy changes desire in still other ways than in inverting the established order of psychic causes and effects. It destroys once and for all the assumption that desire is always directed at objects, explicitly relating it instead to other minds. This follows because our wishes arise from a 'comparison of our condition with that of others'. Thus, as the element of praise and blame is introduced into our evaluation of these artificial passions:

> some, as avarice and envy, are universally condemned; some, as friendship and curiosity, generally praised; but there are others about which the suffrages of the wise are divided, and of which it is doubted, whether they tend to promote the happiness, or increase the miseries of mankind.   (*Rambler*, I, 265)

*   *   *

The most important of this 'ambiguous and disputable' third group of artificial passions is the so-called 'love of fame'. Although this passion is not singled out in Johnson's analysis in *Rambler*, No. 49, it could very well be viewed as a prominent part of a fourth and distinct phase in his anatomy of desire. For the 'love of fame' coincides with the emergence of the heroic mind, an emergence which transforms the desire for objects, already partially modified in the third stage, into a desire for recognition. The examples Johnson cites – Themistocles and Caesar – suggest that the two typological models of this heroic love of fame are the statesman and the warrior–conqueror. It is noteworthy in this connection that 'love' does not figure prominently in Johnson's anatomy of the passions. Johnson's contention in the *Preface to Shakespeare* that modern tragedies have given an exaggerated importance to the passion of love is well-known; love simply lacks the power to become a credible motive for action on the larger scale of public life. For this reason, the encounter between individuals is described in *Rambler*, No. 49, as the result of the desire for praise, honour, or fame. Desire is manifested less in the craving that characterizes love than in the aspiration of the heroic mind for the applause of other minds. Yet this aspiration necessarily forces the individual into competition with other individuals. Lacking this ambition, he would be content to dwell in obscurity. On this basis, the love of fame becomes the foundation of a generalized theory of social conflict.

Throughout his moral writings, Johnson describes this form of social conflict as the universal struggle for emulation. Whatever place it deserves in his anatomy of the passions, it offers a very different explanation of social stratification from that found in the traditional hierarchical world view. In this sense, it differs sharply from what he viewed as Pope and Jenyns's disregard of social conflict in their assumption that social and political ranks are as fixed and static as the hierarchy of creation. Pope and Jenyns's assumption, moreover, involves two separate sources. One of them, made familiar to generations of students through Lovejoy's *The Great of Being*, traces the cosmic and social hierarchies back to a neo-Platonic ladder of life.[3] But this explanation of the scale of being in exclusively idealist categories ignores a second source of the social hierarchy in Aristotle's *Politics*. At the heart of Aristotle's conception of hierarchy is the famous distinction, elaborated in chapter five of *The Politics*, between those who are born masters and those who are born slaves: 'from the hour of their birth, some are marked out for subjects, others for rule' (1254a). According to Aristotle, this hierarchic 'duality . . . exists in living creatures, but not in them only; it originates in the constitution of the universe' (1254a).[4] This suggests that the social hierarchy finds its main theoretical basis not only in Platonic differences of nature but also in different degrees in the same order of being. For it is here that we find the repeatable patterns of the hierarchic relation of master/slave on which civil society rests. In his radical extension of this hierarchic relation to the cosmos as a whole, Aristotle contends that even inanimate objects are governed by 'a ruling principle'.

On the surface, Johnson's conception of a universal contest for superiority seems to resemble Aristotle's belief that hierarchy, not equality, is the natural state of human society. Johnson is at one with Aristotle in believing that men are naturally unequal. Insofar as they are animals, they differ; some are quicker, others slower; some are more cunning, others less so. In addition, Johnson believes that men are also separated by the various artificial goals they pursue – whether these be fashion, accumulation, glory, or whatever else. But below the surface, one can discern fundamental differences between Johnson's view and that of Aristotle. Johnson departs most noticeably from Aristotle's position in his well-known outrage at all institutionalized forms of slavery. But he also differs from Aristotle when he bases the social hierarchy on the artificial competition for prestige and not on natural inequality, as some commentators have mistakenly supposed.[5] The contest for emulation results in social and political distinctions but these distinctions are not necessarily commensurate with natural differences. Indeed, existing political arrangements may actually be inimical to natural eminence. The contention that 'slow rises worth by poverty depressed' is only one instance of an asymmetry between natural talent and material reward that Johnson sees as inherent in all societies.

Another area in which Johnson differs from Aristotle concerns the relationship of master and slave. In Aristotle's account of that relation, the proper functioning of a hierarchical society requires that the master rule the slave in the same way that the intellect governs the passions. When this system of subordination is overthrown, despotism inevitably ensues (*Politics*, 1254b). In Johnson's conception of civil society, by contrast, there is no originating authority, no founding legitimacy by which men spontaneously and automatically give their 'consent' to an institution in which some shall govern and others be governed.[6] Moreover, in the absence of such a legitimating authority and hierarchic principle, the potential for conflict engendered by the contest for superiority can be extended infinitely. Indeed, there is no one, Johnson holds, who does not at some point or other in his life indulge in dreams of absolute power: 'the day is always coming to the servile in which they shall be powerful, to the obscure in which they shall be eminent' (*Idler, Adventurer*, pp. 390–1). Such dreams are usually checked by the *vis inertiae* of custom, but they reveal the extent to which an inclination to accept one's place in the scheme of things is countered, even among the lowliest, by a desire for eminence.[7] In the powerful, this desire is embodied, according to Johnson, in 'ambition'; in the weak, it appears as 'resentment'. In *The Life*, Boswell records a conversation in which Johnson links these two passions to the Aristotelian doctrine of the purgation of pity and fear in tragedy:

> ambition is a noble passion; but by seeing it upon the stage, that a man who is so excessively ambitious as to raise himself by injustice, is punished, we are terrified at the fatal consequences of such a passion. In the same manner a certain degree of resentment is necessary; but if we see that a man carries it too far, we pity the object of it, and are taught to moderate that passion.[8]

But desire can play this role only because it enables individuals to transcend their natural existence. Unlike animals, men desire not merely to survive, to persevere as living beings. They also wish to be recognized as something elevated beyond the level of 'brutes'.[9] But for the attainment of this goal, although it is as important to life as human survival, we cannot rely on the hedonistic calculus of pleasure and pain that marks the first two phases of human desire. In those phases, as in the life of animals, 'pleasure' is the supreme value. The desire for superiority, by contrast, may sometimes manifest itself in the risk of pain, the threat of physical injury and even death.

Thus the universal contest for superiority, like other artificial passions, rests on the suspension of what later philosophers would call the principle of 'utility'. But this suspension only heightens, rather than resolves, the problem of subordination. For the contest for superiority necessarily involves the individual, as we have seen, in a struggle with others, and in its violent form, this struggle can take an aggressive, even malevolent coloration. As Johnson points out in a sermon on strife and envy, 'all violence, beyond the necessity of self-defence, is incited by the desire of humbling the opponents' (*Sermons*, p. 242). When this violence seizes what would have 'been granted to request', it becomes evident that it has been:

> chosen for its own sake, and that the claimant pleases himself, not with the *possession*, but the *power* by which it was gained, and the mortification of him to whom his superiority has not allowed the happiness of choice, but has at once taken from him the honour of keeping, and the credit of resigning.   (*Sermons*, p. 243)

But it is not the unbridled lust of dominion, coupled with a delight in the mortification of others, that provides the primary motivation for the love of fame. What distinguishes the desire for glory from mere natural talent is, above all, the willingness to take risks. Where one individual rises above animal life, able to confront death and not be afraid of failure, another individual will prefer life to fame, oblivion to eminence, prudence to heroic enterprise. This fundamental choice is what distinguishes Johnson's conception of the noble and servile mind from that of Aristotle. He stresses the positive value of competition usually associated with the ideology of a middle class individualism, but his outlook remains aristocratic. Thus while he offers numerous suggestions for the proper 'regulation' of the noble mind, he clearly regards it as vastly superior to the servile consciousness. In *Idler*, No. 57, Johnson describes 'prudence' in largely negative terms. As a virtue, it is not only susceptible to the temptation of idleness; it also 'quenches that ardour of enterprise, by which every thing is done that can claim praise or admiration, and represses that generous temerity which often fails and often succeeds' (*Idler, Adventurer*, p. 178). Johnson cites as an example of prudence in the essay the self-serving caution of Sophron, a character who 'creeps along, neither loved nor hated, neither favoured nor opposed'. Sophron, Johnson writes, 'has never attempted to grow rich for fear of growing poor and has raised no friends for fear of making enemies' (*Idler, Adventurer*, p. 180).

Yet if the noble mind is clearly superior to the vulgar consciousness, it is no more capable than the latter of achieving the plenitude it desires. Indeed, its aspirations are susceptible to the same logic of supplementarity and substitution that characterizes all manifestations of hope. Moreover, the noble mind is limited in an even more fundamental way; it can achieve independence only by making itself dependent on another mind. For solipsism, as we have seen, is always impossible; even within the interior world of voice as presence, the great are only great because they are recognized by the vulgar. And in the exterior world, Johnson contends in *Adventurer*, No. 126, 'greatness is nothing where it is not seen' and 'power nothing where it cannot be felt' (*Idler, Adventurer*, p. 473). This means that power is wholly relative. Indeed, it is even further diminished by the fact that it depends, for its full realization, upon the recognition of beings who are not in themselves independent. As individuals who have lost out in the struggle for emulation or simply opted for 'ease' rather than 'glory', the servile are not in a position to offer their praise sincerely or freely. As a consequence, their applause can only take the form of 'flattery', and flattery is by its very nature self-defeating. Far from granting the recognition that the desire for eminence craves, it offers a satisfaction which seems hollow because it is forced. In *Idler*, No. 100, Johnson gives as an instance of his hollow satisfaction the melancholy of Ortugrul, a rich merchant who, though surrounded by 'flatterers', heard them 'without delight, because he found himself unable to believe them' (*Idler, Adventurer*, p. 305).

*   *   *

Dislodged from its traditional position in the great chain of being, the relation between the powerful and the weak, as Johnson conceives of it, bears a striking resemblance to the unstable dialectic of domination and servitude in Hegel's *Phenomenology of Spirit* (1807). Like Hegel, Johnson insists that the autonomy of the great is actually a heteronomy, that it is constituted only through the mediation of the weak. And Johnson resembles Hegel in emphasizing that the manners of those who are placed in a position of inferiority are really shaped by the wishes of their superiors; the vulgar do not live an independent existence but rather mirror the desires and aversions of the powerful. Glory, fame and honour are thus shown to reside in a reflective consciousness that has no independence of its own. But Johnson also differs from Hegel in at least one important respect. In contrast to Hegel, Johnson immobilizes the dialectic at the moment in which heroic aspiration fails to achieve its expectations. Thus while Johnson parallels Hegel in showing that the servant is really the lord of the lord, he does not mean, like Hegel, to imply by this that the servant achieves independence through the productions of his labour. Johnson was too familiar with the actual deprivations of the working poor in eighteenth-century England to adopt such an easy solution. Instead, Johnson implies that the servant achieves a largely 'ironic' triumph over the master, a triumph which – because it is brought about by the vulnerability of the latter – leaves the servant for the most part still entangled in the coils of material deprivation and his own servile consciousness.

The tensions and ironies which this dialectic of domination and servitude discloses as the unavoidable basis of human relations are more clearly exemplified in Johnson's major poems than anywhere else in his writings. The very notion of an antithesis between freedom and servitude, bravery and cowardice, is an explicit theme in the tradition of heroic tragedy to which *Irene* belongs.[10] In Johnson's sole contribution to this tradition, the theme is subjected to a critique which puts in question its dream of heroic self-sufficiency. In place of an *ethos* that affirms the power of the exemplary hero or patriot, Johnson substitutes a perspective which reveals that power to be as dependent and derived, as the weakness it has supposedly transcended.

In insisting upon the importance of identifying and examining this theme in *Irene*, we run the risk of course of inflating its literary value. Estimating this value has proved in the past to be difficult since Johnson wrote no other plays and since he encountered difficulties in getting *Irene* produced in London.[11] Indeed, the question whether Johnson's tragedy is stage-worthy has perhaps led some commentators to conclude that its vision is boring, inconsistent, or immature.[12] However, one point on which all readers might agree is that *Irene* belongs to a tradition of declamatory tragedy that itself is deemed undramatic by the conventions of the modern theatre. It may thus be worth examining the play's vision in the context of the master–slave dialectic even at the risk of scanting consideration of its stageworthy qualities.

In *Irene*, the struggle for emulation is carried out in the context of a despotism in which there can be only one master, Mohamet, the Emperor of the Turks. Hence, it is not surprising that while Cali Bassa, 'the chief whose wisdom guides the Turkish councils', is 'the highest slave', he is nonetheless 'tir'd of slav'ry' and 'projects' at once the freedom of the Greeks and 'his own' (I. i. 112−14). Cali Bassa is motivated in his schemes against the emperor, not only by his own desire for glory, but also by his awareness of the distrust with which the emperor views him. He knows that if he were existing in a state of real dependency, there would be no strife, for Mahomet might accept him as a slave; but because he knows that Mahomet needs his council, Cali sees himself as an example of the 'unhappy lot of all that shine in courts;/For forc'd compliance, or for zealous virtue,/Still odious to the monarch or the people' (I. ii. 50−2). Just as the emperor is doomed to rely on the advice of courtiers whose flattery he cannot believe, so the slave−courtier must, in turn, depend upon a gratitude he cannot trust. Thus he is no more autonomous in his privileged position than the master upon whom he must rely for his authority. It is not by chance, therefore, that Cali delivers the famous paean to an imaginary land of peace and plenitude:

> If there be any land, as fame reports,
> Where common laws restrain the prince and subject,
> A happy land, where circulating pow'r
> Flows through each member of th'embodied state,
> Sure, not unconscious of the mightly blessing,
> Her grateful sons shine bright with ev'ry virtue;
> Untainted with the lust of innovation,
> Sure all unite to hold her league of rule
> Unbroken as the sacred chain of Nature,
> That links the jarring elements in peace.

> (I. ii. 54−64)

In such a 'happy land', the esteem of the individual flows 'back upon him in its circulation through the whole community' (*Sermons*, 23, 238), rather than being reflected in the capricious visage of a solitary individual. But in the absence of 'common laws' that 'restrain the prince and subject', it becomes impossible to resolve the problem of authority in terms of a 'diffusion of interest' in which each 'member of the embodied state' can spontaneously and willingly sublimate his self-love in the interest of the whole and willingly choose his place in the sacred 'chain of Nature'. Instead, the problem of authority must be posed in terms of a 'contraction' of interest which, inasmuch as it can perceive only its own ends, may be reflected in the 'ambition' of a Cali or in the 'resentment' of his chief lieutenant Abdalla (IV. viii. 9). Constituting the dual sources of fear and pity in tragedy, these passions also insure a permanent source of division in the body politic. It is perhaps for this reason that Cali's dream of a 'happy land' is never presented in *Irene* as normative – as the theoretical basis for a society resting on the principles of plenitude and hierarchy – but only as the momentary fancy of a beleaguered courtier.[13]

It is precisely the continuing presence of such a source of division – a division that fosters the extremes of repression and rebellion – that distinguishes the Turkish despotism of *Irene* from the hierarchical society envisioned in Cali's speech. This applies with particular cogency to the case of the emperor. For power confers on Mahomet the capacity to compel others to do his bidding, but it does not necessarily confer the repute or authority of wisdom. The latter may devolve on the courtier slave who is then able to transform his servitude into mastery, not through labour in the Hegelian sense, but through a plebeian shrewdness, prudence and statecraft. This is nowhere more apparent than in Cali Bassa's account of his intervention, prior to the action in the succession to the throne:

> When unsuccessful wars, and civil factions
> Embroil'd the Turkish State – our Sultan's father
> Great Amurath, at my request, forsook
> The cloister'd ease, resum'd the tott'ring throne,
> And snatch'd the reins of abdicated pow'r
> From giddy Mohamet's unskilful hand.

(I. ii. 41–6)

The title of emperor or king is thus no guarantee of authority in a despotism, for it is not always synonymous with power. The alienation of power from authority, to be sure, may prove to be only temporary. But the consequences of that alienation – a loss of faith in the notions of the emperor's presence, of his sacred role as font of plenitude – is likely to be more far-reaching. In the absence of real authority, the emperor's power becomes visible in *Irene* as something achieved, a product of the same 'Ambition' that fires Cali. Mohamet's 'love' for the Greek slave, Irene, is evidence of a similar ontological slippage. A tyrant who is enslaved to his own passions, Mohamet is in effect no different from his enslaved subjects.

The diminution of sacred authority into secular ambition is rooted, then, in the nature of a specific political system, and the relevant aspect of this system is the displacement of rational values by the wholly unpredictable power of the passions. Thus ambition is opposed in *Irene*, not to obedience or loyalty, but to the creaturely passion of fear that Irene, in her speech on ambition to her companion in bondage, Aspasia, attributes to 'vulgar souls':

> Ambition is the stamp, impress'd by Heav'n
> To mark the noblest minds, with active heat
> Inform'd they mount the precipice of pow'r,
> Grasp at command, and tow'r in quest of empire;
> While vulgar souls, compassionate their cares,
> Gaze at their height and tremble at their danger.

(III. viii. 111–16)

Irene is of course conforming to a venerable tradition of heroic drama when she insists in this passage that the servitude to the body embodied in the unheroic fear of 'danger' is the chief feature distinguishing vulgar souls from noble minds. The vulgar are the slaves, not of their superiors, but of life; they are servile because they flee in the face of danger, preferring bondage to liberty and the prospect of injury or death. An example of this kind of servile consciousness in *Irene* is the 'gasping coward', Menodorus who, 'still fond of life', betrays Cali's schemes to Mohamet's minion, Mustapha. Thus portrayed, Menodorus becomes little more than a simple foil. But Johnson obviates this negative stereotype by emphasizing the link between the fondness for life and the 'unprofitable, peaceful, female virtues' that Cali condemns in an early exchange with Abdalla. The passive, feminine character of corporeal existence is manifested in the 'bonds of nature' that link 'brother', 'friend' and 'father'. Heroic aspiration, as Cali defines it, resides precisely in the domination of these passions by ambition:

> This sov'reign passion, scornful of restraint,
> Ev'n from the birth affects supreme command,
> Swells in the breast, and with resistless force,
> O'erbears each gentler motion of the mind.

(III. i. 18–21)

If one were to read only these set speeches, one might easily assume that the play proper was dramatizing, in conventional heroic terms, the ascendancy of this 'sov'reign passion' of ambition over the creaturely passions of the body. Almost the opposite, however, is true. Although the figure of woman becomes identified in Cali's speech with what is passive, impotent, and fearful, this state of impotence proves itself to be a motive force for action in *Irene*. In striking contrast to Hegel, Johnson finds fear itself to be a source of heroic ambition. This is most apparent in his portrayal of Irene herself. Though Irene believes that she is undertaking a heroic action 'in quest of empire', her speech to Aspasia proves to be a parody of heroic aspiration. For her ambition and apostacy have, from the beginning of the play, been shaped by the very 'cares' she scorns. By pointing to the motives on which they are actually based, Aspasia hints at the predicament they were meant to conceal:

> Well may'st thou hide in labyrinths of sound
> The cause that shrinks from Reason's powerful voice.
> Stoop from thy flight, trace back th'entangled thought,
> And set the glitt'ring fallacy in view.

> (III. viii. 125 – 8)

The ambition of Irene is thus an impure passion that seeks to utilize power as a means of escaping from a servile condition. In a crisis, it is utterly unable to confront the reality of pain and death to which the schemes of Abdalla has condemned her. Irene's final pleas for life are a measure of the degree to which her ambition has all along been sustained by the very fear it was meant to suppress:

> O name not death! Distraction and amazement,
> Horror and agony are in that sound!
> Let me but live, heap woes upon me,
> Hide me with murd'rers in the dungeon's gloom,
> Send me to wander on some pathless shore,
> Let Shame and hooting Infamy pursue me,
> Let Slav'ry harrass, and let Hunger gripe.

> (V. ix. 31 – 7)

The conduct of Irene from the beginning of the play to her downfall thus enacts with perfect logic the necessary conclusion of her aspiration to overcome her servitude. Of all the aspects of the play it is the least dependent upon the historical subject matter, the least directly traceable in its origins to Johnson's source, Knolles's *Generall Historie of the Turkes* (1603). In Knolles, Irene remains largely inert and undeveloped as a character. Johnson, expanding her role in the play, not only heightens dramatic intensity, but also makes her tragic death spring directly from her behaviour. This outcome does not merely reflect Johnson's hatred of apostacy, or his scorn of plebeian ambition. Rather it provides a concrete contrast to the fate of Aspasia, capable of exemplifying an alternative yet logically necessary course of action. In Johnson's moral dialectic, a slave whose life and body are endangered may, like Aspasia, adopt a posture of Christian resignation or indifference. Or, a slave may become ambitious and seek, like Cali or Irene, to achieve power through guile or violence. Needless to say, the latter alternative is incompatible with the passion of 'love', for a person cannot be both a slave and 'in love' with his master at the same time. Yet this is what Mohamet clearly demands of Irene, and his inevitable disillusionment provides the motivation for her final humiliation and death.[14]

It might be argued that Irene's humiliation at the end of the play merely confirms Cali Bassa's conception of 'female virtues'. Nothing could be more conventional than this portrayal of the terrors to which women are subject. Yet it is noteworthy that Aspasia describes these terrors as a fabrication. With great clarity, she argues that they are a social construct:

> The weakness we lament, our selves create,
> Instructed from our infant years to court
> With counterfeited fears the aid of man;
> We learn to shudder at the rustling breeze,
> Start at the light, and tremble in the dark;
> Till Affectation rip'ning to belief,
> And Folly, frighted at her own chimeras,
> Habitual cowardice usurps the soul.

> (II. i. 26–33)

Irene attributes her own exemption from the 'habitual cowardice' of her sex to the 'precepts' of her lover, the 'pious hero' Demetrius (II. i. 38–41).[15] But if men are not subject to this habitual cowardice, they are nonethelesss susceptible to a creaturely weakness that undermines and ultimately subverts their aspirations. In *Irene*, this weakness is described as a guilty fear, a kind of self-division or *dédoublement* which, as the

concomitant of 'treason' and apostacy, 'intimidates the brave' and
'degrades the great'. Cali Bassa's guilty fear is manifested to the world in
the 'disorder'd air' that betrays 'the wild emotions of his mind'. Mustapha
describes this fear as the consequence of treason, a kind of ironic fall from
greatness:

> See Cali, dread of kings, and pride of armies,
> By treason levell'd with the dregs of men.
> Ere guilty fear depress'd the hoary chief,
> An angry murmur, a rebellious frown,
> Had stretch'd the fiery boaster in the grave.

> (IV. viii. 16 – 20).

The extent to which Cali illustrates how the ambitious slave is caught up,
against his own will, by his guilt can be seen in his disclosure under
torture of his scheme to murder Mohamet in Irene's bridal chamber
(V. xi. 12 – 16). Cali's contempt for the creaturely passion of fear thus
proves to be no defence against its actual workings. His heroic ambition is
subverted by his own plebeian weakness.

   This fundamental limitation to heroic aspiration has consequences that
are political as well as ethical. If ambition is attended by fear, there is
nothing that can distinguish the slave from the master. The confusion
engendered by this merging of roles extends throughout the Turkish
realm of *Irene*. If Irene has degraded love by placing it in the service of
ambition, so Mohamet has debased ambition by placing it in the service
of love. The effect on both characters is virtually identical. Thus
Mohamet, fully as much as Irene or Cali, suffers from 'the anguish' which
'racks the guilty breast,/In pow'r dependent, in success deprest' (*Prol.* 11 –
12). Wretched in love, abject in glory, Mohamet is a simple instance of a
master who finds himself in bondage to the slaves he has conquered. In
the simplest sense, Mohamet's bondage is shown in the way he is forced
to rely upon intermediaries. Though his minions, Hasan and Caraza,
faithfully carry out his orders to execute Irene, their zeal illustrates the fact
that the master who can only govern through prostetic agents is reduced,
at crucial moments, to serving them. It might be supposed that Johnson is
merely conforming to a time-worn dramatic convention in making
Mohamet act through agents. But Johnson reveals Mohamet's dependency in a more complex way. Though he is obviously a 'slave' to his
passion for Irene, his power makes it impossible to accept her love even
when it is profferred. For Mohamet's passion embodies the hopeless

desire, not only to 'conquer' and possess the beloved, but also to be loved by one who is still free. In a striking passage, Mohamet suffers from the anguished awareness that:

> Ambition only gave her to my arms,
> By reason not convinc'd, nor won by love,
> Ambition was her crime, but meaner folly
> Dooms me to loath at once, and doat on falshood,
> and idolize th'apostate I contemn.

(IV. vii. 7 – 11).

Here is a prefiguring not only of Hegel's dialectic of domination and servitude in the *Phenomenology of Spirit* but of Jean Paul Sartre's application of this dialectic to the passion of love in *Being and Nothingness* (III. i. 4). The all-important consequence of Mohamet's realization that his love can never be reciprocated is that it swiftly disintegrates into the unstable and complex amalgam of loathing and doting that provides the dual motives for his contradictory actions in the play's final scenes.

Mohamet's bitter disillusionment also helps to explain why Johnson came later to question the importance attached to love in the tragedies of the seventeenth and eighteenth centuries. For Mohamet's love proves, in the last analysis, to be as much of a compound as the ambition it supposedly displaces. Far from being a pure passion that has little or nothing to do with the exercise of worldly power, love is shown to be associated with that power in a way that robs it of its dignity and putative independence from material considerations. Mohamet's power is not an object of admiration, as is the 'virtue' of Demetrius, but it is the means by which he appropriates the other, which the virtue of Demetrius cannot.

*   *   *

At first glance, the patriotism of Demetrius, Leontius, and Aspasia might appear to offer a normative contrast to the unbridled ambition of Cali, Irene, and Mohamet. There is no question that the 'patriotism' and Christian virtue of the Greeks is being opposed in Johnson's tragedy to the ambition and vice of the Turks. In contrast to Turkey, Greece is the temporal homeland of 'liberty' in *Irene*, the privileged site of Christian patriotism, the effort to incarnate heroic virtue in history. Much like Egypt and Rome in Dryden's *All for Love*, Turkey and Greece become the mythical sites of a system of oppositions: tyranny vs. liberty, greatness vs. goodness, guilt vs. innocence, corruption vs. virtue.

Upon closer inspection, however, the contrast between Turkey and Greece turns out to be less profound than it appears.[16] Even before the action of the tragedy, the city of virtue and Christianity, the nation of patriot heroes, has awakened from its dream of peace and plenitude to discover itself in subjugation to Turkey and the reign of tyrants. In the wake of the catastrophe, retold in the opening scene by the Grecian heroes, Demetrius and Leontius, the privileged place given to liberty and patriotic virtue in the system of oppositions which ostensibly governs the play is subjected to considerable strain. Peace, the growth of commerce, and the ensuing 'lust for gold', have all led to neglect, a loss of energy, and a false sense of security (I. i. 2–35). When the norms embodied in Cali Bassa's 'happy land' speech are compared to the devastation described by Demetrius, they seem Utopian. Insofar as they were embodied in Greece, they are shown to have been undermined by decadence and degeneration long before the Turkish invasion. When Demetrius depicts the signs of decay in his homeland, he portrays these signs as a separation from origins:

> A thousand horrid prodigies foretold it.
> A feeble government, eluded laws,
> A factious populace, luxurious nobles,
> And all the maladies of sinking states.

> (I. i. 34–9)

Virtually all of the characteristics that define the figure of decline and fall in eighteenth-century political thought are recorded here: an evasion of laws, a fall from unity into division, an enervation of government, the degeneration of the nobility, and the gradual transformation of freedom into servitude.[17] Leading inevitably to the external catastrophe of the Turkish invasion, this model of political degeneration is already an internal catastrophe.

The destruction of Greece and the capture of its leaders necessarily places a severe strain on the credibility of its patriotic values. The 'reality' of Turkish power always remains on the horizon of *Irene*, and the Greek heroes, necessarily reduced to being the leaders of a band of captives, remain circumscribed within this horizon. In the absence of personal 'liberty', Leontius and Demetrius are forced to rely on the plebeian Machiavellianism of Cali and Abdalla. And this Machiavellianism, which replaces the heroic ideal of open combat with plans for an assassination and 'speedy flight', can hardly be expected to lead to a restoration of the lost *patria*. As Leontius pointedly asks,

> But what avails
> So small a force? or why should Cali fly?
> Or how can Cali's flight restore our country?

> (I. i. 137–9)

What Cali proposes to Demetrius and Leontius in the course of the opening scene is, in effect, a personal flight which, relying upon the 'confusion' of the Turks, is supposed to lead to the establishment of an 'easy Throne' and the 'retreat' of the Turks 'from European shores' (I. ii. 74–82). The gulf separating the plebeian values underlying this scheme and the Grecian ideals can be seen in the exchange of Leontius, Cali, and Demetrius over the dangers of the plan. Asked by Leontius, 'What has the wretch that has surviv'd his country,/His friends, his liberty, to hazard?' Cali replies, 'Life'. The irony of Demetrius's response, 'Th'inestimable privilege of breathing!/Important hazard! What's that airy bubble/When weigh'd with Greece, with virtue, with Aspasia' (I. ii. 133–8), underscores the commitment of the Grecian heroes to a code based on honour and the neglect of personal safety. Yet in spite of this commitment, they are still forced to rely on the machinations and betrayals which are the essence of Cali's scheme and which eventually doom their aspirations to failure.

Yet if the Greeks fail to achieve their patriotic aspirations, they could still be said to demonstrate their heroic values as individuals. Demetrius, Leontius, and Aspasia appear to evince a personal courage and virtue that distinguished them from Irene and the Turks. But even when the Grecian characters are examined from a perspective that is ethical rather than political, their actions display some of the same features that undermine the ambition of the Turks. The scene in which Demetrius and Leontius must decide who will fight and who will stay poses a test for the patriotic ideal, which depends for its success on the willingness of individuals to subordinate their private interests to the public good. Leontius points out the significance of this test when he asks 'Why has thy choice then pointed out Leontius/Unfit to share this night's illustrious toils?/To wait remote from action, and from honour' (IV. iii. 20–2). It is not a coincidence that Demetrius's reply, 'Can brave Leontius be the slave of glory?' poses the question of patriotism in terms of mastery and servitude. For the patriotic

ideal shares the same ideal of self-mastery as the 'ambition' of the Turks.
When Leontius pointedly asks Demetrius, if he will 'head then the troop
upon the shore/While I destroy th'oppressor of mankind', Demetrius's
reply displays the very same submission to 'glory' that he finds in
Demetrius's question:

> What can'st thou boast superiour to Demetrius?
> Ask to whose sword echo through the shouting field;
> Demand whose force yon Turkish heroes dread,
> The shudd'ring camp shall murmur out Demetrius.

> (IV iii. 38–42)

It is only when in order to break the vicious circle of master vs. master that
Leontius offers to subordinate his aspiration, to serve as second to
Demetrius, that the conflict is resolved. The scene – which has been
unjustly criticized for its awkwardness – serves to underscore the
contradiction inherent within the supposedly patriotic virtues of the
Greeks.[18] If Irene embodies the 'greed' which undermines the Greek
experience at once extreme, Demetrius emobodies the 'pride' which
corrupts it at another.

Moreover, the catastrophic failure of Cali's plans necessarily becomes a
personal failure for the Greek protagonists. Patriotic virtue cannot be
preseved in an unholy alliance with treason. Aspasia points out the ethical
limitations of a scheme based on the subordination of noble ends to base
means:

> Think how the sov'reign arbiter of kingdoms
> Detests thy false associate's black designs,
> And frowns on perjury, revenge, and murder.
> Embark'd with treason on the seas of fate,
> When Heav'n shall bid the swelling billows rage,
> And point vindictive lightnings at rebellion,
> Will not the patriot share the traytor's danger?

> (IV. i. 56–62)[19]

Aspasia's wish 'Oh could thy hand unaided free thy country,/Nor
mingled guilt pollute the sacred cause!' (IV. i. 63–4) reveals the contamination that has infected Grecian values. Though her scruples are dismissed
as the 'phantoms of a woman's fear', they point up the conflict between an

ethic of liberty and an ethic of Machiavellian treason. For the 'ambitious slave', who is motivated by fear or resentment, all the means to the desired end, are good. But in thus subordinating ends to means, the slave reveals that he lacks the essential element of the internal dialectic of mastery, the conquest of self. His belief is that he can become master, easily, by outwitting or ambushing his opponent rather than by risking his own life.

The central drama of the Greeks in *Irene*, then, is the corruption of virtue which results from its association with vice. In the absence of civil liberty and civic unity, patriotism becomes divorced from integrity, the real from the ideal, seeming from being. Thus if the 'patriot' shares 'the traitor's danger', he also shares in the pollution which corrupts his cause. To rely on schemes and 'Conjectures' becomes a way of choosing liberty through the mediation of the 'other'. It is the main example of *mauvaise foi* in *Irene*. Refusing recourse to direct action, Demetrius and Leontius have, in effect, placed themselves in the hands of slaves eager to prove their own ascendancy. Appropriately enough, it is Aspasia, who had already recognized most clearly the corruption inherent in the Greek project, who also proves willing to take on the 'guilt' for its catastrophic failure (V. iii. 10–13). Demetrius's injunction to Aspasia to 'devolve . . . thy sorrows on the wretch,/Whose fear, or rage, or treachery betray'd us' (V. iii. 16–17) should not disguise the contamination of the Greek design by its alliance with the 'fraud' and 'falshood' of Cali and Abdalla. Though responsibility for this 'contamination' is to some extent displaced by its ascription to the treachery of the fiery Abdalla, it is not eliminated. The inconclusive encounter of Demetrius and Abdalla, and the subsequent escape of Aspasia and Demetrius through the passive assistance of Irene prove only the blessings which reason and virtue, assisted by Divine Providence, can bestow on the individual, Patriotism, like ambition, proves inadequate to deal with the perils which beset it.

It is precisely the presence of such a sense of deeply felt and personal corruption – a corruption that admits no real apology or defence – that should warn us against dismissing *Irene* too easily as one more instance of a 'chill' neo-classical tragedy or of condemning it too quickly in the light of our expectations of what is dramatic. Because of the immense gulf that separates these expectations from the conventions of Johnson's theatre, *Irene* may possess a thematic significance that has been too easily overlooked in the twentieth century. Despite appearances to the contrary, Johnson may have been as eager to have *Irene* represent something – even, in some special sense, to have it represent 'nature' – as the most straightforward realist dramatist of a later era, and the dialectic of domination and servitude may have given him a theme beyond the reach of custom and appearance to represent in a meaningful way. Thus, *Irene* might be said to have its genesis in a declamatory, rhetorical version of the representational art that Johnson praises so highly in his *Preface to Shakespeare*.

4

The Decline of the Heroic: from *London* to *The Vanity of Human Wishes*

Johnson devoted other poems besides *Irene* to the dialectic of domination and servitude. *London* and *The Vanity of Human Wishes*, his adaptations of Juvenal's third and tenth satires, published in 1738 and 1748, are similarly preoccupied with antitheses of hope and fear, bravery and cunning, heroic and servile states of consciousness. But where *London* adopts a position complementary to that of *Irene*, *The Vanity of Human Wishes* represents a decided change in point of view. This change is most apparent in its attitude toward the illusion of self-sufficiency that is at the centre of the heroic idea. In the Grecian world of *Irene* and the English world of *London*, this illusion is very much alive, enshrined in the contrast between a corrupt and servile present and a heroic past. Even though this past is clearly marked as past, as vanished, it nonetheless embodies the still-living myth of a founding simplicity, self-mastery, and self-identity. But in *The Vanity of Human Wishes*, this distinction between corrupt present and heroic past virtually collapses, as the will-to-power becomes inscribed in a universalizing syntax of servile human wishes in which there is neither origin nor end.

*   *   *

Composed during the same period of his career as *Irene*, Johnson's first major poem, *London, A Poem in Imitation of the Third Satire of Juvenal* (1738) displays a political perspective that is markedly similar. Adopting the rhetoric and 'country ideology' of the patriot opposition that flourished during the 1730s, *London* portrays an England teetering on the brink of possible destruction. Once again, the nation is threatened by a foreign power (Turkey in *Irene*; Spain in *London*); and once again, avarice is seen as a major source of corruption. Though the argument, thus formulated, may well seem 'jejune' when compared to that of *The Vanity*

47

of Human Wishes, London is much more than a catalogue of 'the meerest commonplaces of opposition propaganda'.[1] Organized around a conventional myth of England's precipitous decline, *London* dramatizes this decline in terms of a coherent vision – the vision of an invasion of what is integral, self-sufficient, and self-identical by what is alien, parasitic and external to it. While Johnson conforms to the essentials of Juvenal's third satire in his elaboration of this vision, his choice of texts need not be thought of as academic: it is wholly consistent with the 'country ideology' that informs his satire in the poem.

At the core of this ideology is the conviction that the movement from England's glorious past to its degenerate present is at the same time a movement from domination to servitude. 'Brittania's glories' are largely defined in terms of allusions to popular national heroes – Eliza (ll. 23–8), Edward II (ll. 99–102), Henry V (ll. 117–23) and Alfred (ll. 248–53). These larger-than-life-size figures evoke the image of an England that was once single and undivided, a 'land of heroes and saints' who were identical with one another in their virtue and patriotism (l. 100). The history of England, as embodied in the harangue of Johnson's Juvenalian *persona*, Thales, is the story of the progressive degradation of the populace from this ideal of heroic self-sufficiency to the reality of its present servitude and self-division. Thales explicitly attributes this degradation to the inundation of London by 'the dregs of each corrupted state' (l. 95).

The decline of England from its heroic past has ethical as well as political implications. The corruption of the nation has as its counterpart the subversion of the manly struggle for emulation and superiority. In the nation's golden age, this struggle was governed by 'justice' – that is, by the equitable distribution of rewards and punishments. 'Worth' and 'virtue' were given their due, not for what they appeared to be, but for what they actually were. But in the degenerate present, power and justice have become alienated from one another. Located on the margin of society, Thales appears to be an apt symbol of this alienation. Lacking even 'the cheap reward of empty praise', Thales seems to exemplify the dispossession of true worth and 'surly virtue'.

As such, Thales might well appear to conform to the set of expectations by which the Juvenalian satiric *persona* is conventionally defined. Yet because of the intensely polemical nature of Johnson's political rhetoric, it is not easy to describe his attitude toward his protagonist. For while Thales appears to be a typical Juvenalian patriot hero, he is a curiously non-virile one – almost a caricature of Leontius or Demetrius. His departure for Scotland, for example, represents not a return to a heroic world of primal fulness but to what the narrator who introduces Thales's speech describes as a pre-heroic world of vacuity and desolation (ll. 7–12); and Thales, much like the sycophants he derides, fails to distinguish the harsh reality of this world from the 'pleasing dream' of a 'purer air'. It is a sign of his lack of manliness that he should transform this 'pleasing dream' into an English country estate. Portraying this estate as the 'deserted seat' of a 'hireling Senator', Thales organizes his description around a number of conventional oppositions: country/ city; health/sickness; harmony/division; simplicity/complexity; frugality/luxury (ll. 210–23). Yet insofar as this estate is the product of wealth and contrivance, it is subject, as D. V. Boyd and T. K. Wharton observe, to the same charge of luxury and artfulness that Thales aims at the contrasting London world of 'park and play' (l. 210).[2] Moreover, even if this estate is accepted on Thales's own terms, as a kind of edenic landscape garden, it is still delimited by the unheroic values of personal ease and security. As an indigent scholar, Thales could only live in such a setting as a parasite, the very type of the corruption he castigates in London. Compared with the legendary characters he invokes, Thales is thus a diminished figure, more of a hero-worshipper than hero. Even in his 'surly virtue', Thales seems to embody something of the servility and puzzled defeat he attributes to his contemporaries.

Thales's passivity is of course consistent with his view that England has become a nation of slaves. Resembling the escape of Demetrius and Aspasia from Turkey, his self-imposed exile is more of a pastoral retreat than a heroic triumph. As such it is less a remedy than a symptom – a flight from corruption that simply reverses but does not forestall the process by which England has been invaded from the outside, its unity transformed into division, its strength into weakness. For the neglect of true worth is also the appropriation of its rightful place in society by the dispossessed – by those who were originally beyond its precincts. Thales's horror at this process of appropriation reaches a climax in his ironic apostrophe to the 'cheated nation's happy fav'rites':

> Mark whom the great caress, who frown on me!
> London! the needy villain's gen'ral home,
> The common shore of Paris and Rome;
> With eager thirst, by folly or by fate,
> Sucks in the dregs of each corrupted state.
> Forgive my transports on a theme like this,
> I cannot bear a French metropolis.

(ll. 92–8)

This passage points to an evil which is wholly alien, parasitic and exterior. What is especially pernicious about this evil is that it has introduced a servile consciousness where none existed before. At the heart of Thales's 'pleasing dream' of England's heroic past is the conviction that it was once a nation of masters, a community of equals. By contrast, it has now become a nation of masters and slaves. Where the noble consciousness of the English freeholder once dwelled in the solitary splendour of his self-approving virtue, it now finds itself reflected in the base consciousness of the outsider. The most dangerous aspect of this base consciousness is not its moral laxity but its obsequiousness. Implicitly opposed to the 'surly grace' of the English patriot of old, the servility of the immigrant is manifested, above all, in his desire to please:

> Obsequiousness, artful, voluble, and gay,
> On Briton's fond credulity they prey.
> No gainful trade their industry can 'scape,
> They sing, they dance, clean shoes, or cure a clap;
> All sciences a fasting Monsieur knows,
> And bid him go to hell, to hell he goes.

(ll. 111–16)

This rhetoric of flattery is of course based on an obvious talent for mimicry:

> Wit, brav'ry, worth, his lavish tongue bestows;
> In ev'ry face, a thousand graces shine,
> From ev'ry tongue flows harmony divine.

(ll. 126–8)

Unlike the surly virtue of the native-born Englishman, the histrionic gifts of foreigners do not go unrecognized; we are told that 'with justice, this discerning age/Admires their wond'rous talents for the stage' (ll. 133–4). Such 'talents', moreover, are beyond the capacity of native Englishmen to emulate. Indeed, Thales detects in their feeble efforts at mimicry something of the awkwardness and strain that, for him, characterizes the plight of 'starving worth' in contemporary London. Unlike the Frenchman who, born a 'parasite', is 'to his int'rest true, where 'ere he goes', the English patriot is divided. Self-consciously representing himself to himself in the theatre of his mind, he futilely attempts to restore what is already lost. Hence his efforts to acquire the 'arts' of masquerade are 'in vain'; all they really accomplish is to 'strain out with fault'ring diffidence a lye,/ And get a kick for aukward flattery' (ll. 129−31).

Yet if 'rugged natives' fail at flattery, their deficiency is more than made up for by the skill of foreigners who, in their efforts to cater to the whims of their hosts, transform their strangeness into specular familiarity. Through mimicry, which is the 'supple' Gaul's substitute for the power he lacks, he becomes a mirror in which the Londoner finds his image reflected:

> Well may they venture on the mimick's art,
> Who play from morn to night a borrow'd part;
> Practis'd their master's notions to embrace,
> Repeat his maxims, and reflect his face;
> With ev'ry wild absurdity comply,
> And view each object with another's eye;
> To shake with laughter ere the jest they hear,
> To pour at will the counterfeited tear;
> And as their patron hints the cold or heat,
> To shake in dog-days, in December sweat.
>
> (ll. 134−43)

Such specularity is dangerous precisely because it goes unrecognized. Substituting fancy for judgment, identity for difference, laxity for moral rigour, it gradually leads to a situation in which the self becomes dependent on the other for his sense of self-esteem:

> For arts like these preferr'd, admir'd, carest,
> They first invade your table, then your breast;
> Explore your secrets with insidious art,
> Watch the weak hour, and ransack all the heart;
> Then soon your ill-placed confidence repay,
> Commence your lords, and govern or betray.
>
> (ll. 152−7)

It is precisely at this point that the master becomes the slave of the slave in the Hegelian sense, and the slave becomes the master of the master. The 'supple' Gaul's subversive and compliant effort imperceptibly transforms servitude into domination: while the English lord attains only a transitory enjoyment, the foreign servant obtains the secrets and confidences whi h enable him to 'govern or betray'.

Johnson's visio n of a triumph of servants over masters is adapted from Juvenal's third sat, e, but his most important alteration – the shift from a pastoral to a heroic past – heightens the ideological aspect of this triumph.[3] As power passes from master to servant, it becomes available in effect to whomever can pull the strings of influence or purse. When the importance of courage or virtue as a source of identity begins to wane, ascriptive status emerges as a commodity and with this emergence comes a change in the very nature of identity itself. It becomes visible as something bought or sold, a human product contingent upon wealth, influence, or service:

> But here more slow, where all are slaves to gold,
> Where looks are merchandise, and smiles are sold,
> Where won by bribes, by flatteries implor'd,
> The groom retains the favours of the lord.

> (ll. 177–81)

Not surprisingly, poverty becomes the visible emblem of alienation: 'All crimes are safe, but hated poverty./This, only this, here rigid law pursues' (ll. 159–61).

Thales's comments on poverty are marked by the same ambivalence that characterized his utterances on his departure from London. On the one hand, he denounces the corruption that occurs when all become 'slaves to gold'. Voicing the typical concerns of the patriot opposition, he attacks a possessive market society in which a person's value becomes his commodity and his price in the market-place. On the other hand, in deploring an age in which *'Slow rises worth, by poverty deprest'* (1.76), Thales implicitly sustains the view that 'worth' is no longer its own reward but has become a means to an end. What formerly confirmed an absolute value based on the self-esteem of the master has now lost its savour and is worthless unless marketable. Thales is so much a part of his age that he laments not only the alienation of true worth but its failure to achieve the material reward it deserves in a nation of slaves. To add insult to injury, the reward of poverty in such a society is not only neglect, Thales laments, but also scorn:

> This, only this, provokes the snarling muse;
> The sober trader at a tatter'd cloak,

Wakes from his dream, and labours for a joke;
With brisker air the silken courtiers gaze,
And turn the varied taunt a thousand ways.
Of all the griefs that harrass the distrest,
Sure the most bitter is a scornful jest;
Fate never wounds more deep the gen'rous heart,
Than when a blockhead's insult points the dart.

(ll. 161–9)

If flattery is the medium by which the rich and powerful are acknowledged in a community of slaves, scorn is the means by which the poor are
abused. Yet inasmuch as they are equally hungry for recognition, it is not
surprising that 'the gen'rous heart' is never more deeply wounded 'than
when a blockhead's insult points the dart'.

The emergence of a market society thus parallels fairly closely
England's decline from its former glories: the dual movement is
underscored by moral–social images of rising and falling, of conflagration
and collapsing houses. These images draw attention to the moral decay as
well as physical danger of the city. Both are a reflection of the transition
from heroic self-sufficiency to servitude, a transition in which the
populace, as slaves to gold, have lost the courage that once distinguished
masters from servants. No one is exempt from this servile outlook, not
even the 'lords of the street' who, as outsiders in an acquisitive society,
might appear to have escaped its values:

Yet ev'n these heroes, mischievously gay,
Lords of the street, and terrors of the way;
Flush'd as they are with folly, youth, and wine,
Their prudent insults to the poor confine;
Afar they mark the flambeau's bright approach,
And shun the shining train, and golden coach.

(ll. 230–5)

Like parasites and actors, those other agents of corruption and division,
the 'lords of the street' are at once alienating and alienated. Flushed 'as
they are with folly, youth, and wine', they nonetheless participate in the
hierarchical yet servile chain of slaves that marks the new society. In place
of a community of free, self-sufficient individuals linked by ties of
mutuality and respect, *London* projects a dispersive world in which

flattery and insults, linking one to those above and below, fail to preserve the bonds that hold men and women together in a city of heroes. The 'lords of the street' participate in this servile world by confining their 'prudent insults' to the 'poor', but in so doing only confirm the real alienation and dispossession that characterizes this world.

*   *   *

Johnson's assault in *London* on the new hierarchy of slaves hinges on what is an essentially metaphysical opposition of the autonomous to the parasitic, of the native to the foreign. The positive side of this opposition – the notion of a humanist prince at the apex of society – ensures plenitude by sustaining the vision of a nation of public-spirited patriots and heroes. In other words, the opposition is implicitly compatible with the theological principles attacked in the review of Jenyns's *Inquiry*: continuity, hierarchy, closure. But in *The Vanity of Human Wishes* the hierarchical contrast between self-sufficiency and dependency virtually disappears in the speaker's denunciation of the whole idea of epic virtue. Yet while the theme of this denunciation is the vanity of all human aspirations, it is not allowed to become the pretext for a displacement of heroic glory by the contrasting virtues of peace, temperance, wisdom, and heroic resignation. The result of such a reversal might be the reconstitution, in a new guise, of the same antithesis of autonomy and dependency that governed *London*. In *The Vanity of Human Wishes*, this very antithesis becomes tenuous, as the quest for self-sufficiency is revealed to rely on the wishes and fears it supposedly overcomes.[4]

To understand how this opposition is undone in *The Vanity of Human Wishes*, we need to look briefly at the way in which it is presented in the poem. While the dichotomy between the heroic and the servile is not nearly as prominent in *The Vanity of Human Wishes* as it is in *London*, it is nonetheless clearly visible in a series of antitheses that organize the speaker's polemic: hope/fear; desire/hate; the bold/the knowing; the warrior/the scholar; the conqueror/the suppliant. These antitheses depend upon the same hierarchical values which inform *Irene* and *London*. The conqueror is distinguished from the suppliant by the courage which enables 'Swedish Charles' to become the 'unconquer'd lord of pleasure and of pain' (l. 196). Moreover, *The Vanity of Human Wishes* can by no stretch of the imagination be said to abandon the rhetoric and vocabulary of the patriot opposition of the 1730s. The speaker's indictment of the 'supple tribes' who 'repress their patriot throats/And ask no question but the price of votes' (ll. 95–6) is a direct continuation of Thales's critique of those 'whom pensions can incite/To vote a patriot black, a courtier white' (ll. 51–2). But the effect of the speaker's indictment is not to reassert the metaphysical oppositions that prompted Thales's search through English history and legend for examples of patriotic heroism. The desires, gestures, and actions that established the ascendency of the heroic ideal in *London* now also displace that ideal from its position of primacy. And yet conversely, as soon as the mark of servile consciousness, flattery, turns into its opposite, hatred, the speaker's irony begins to waver and heroic consciousness starts to regain the upper hand. The final effect of this pattern of reversibility is that the inseparable figures of master and slave resume their perpetual struggle for dominance at the heart of human existence.

In one sense, to be sure, the heroic *ethos* of *London* appears to escape irony in *The Vanity of Human Wishes*. In his choice of exempla, the speaker avoids the patriot leaders who provided the historical subject matter for Thales's celebration of England's glorious past in *London*. In a shift reminiscent of the movement from *Henry V* to *Henry VIII* in the Shakespearean canon, Eliza, bluff Prince Hal, Edward II, and Alfred are replaced by Laud, Wolsey, Villiers, Wentworth, Harley, and Hyde. As instances of the courtier-statesman, these native figures are counterposed to the foreign examples of the warrior-prince: Charles XII of Sweden, Charles Albert of Bavaria, and Xerxes. Yet it is uncertain whether Johnson considered the English patriot king an apt model for debunking in *The Vanity of Human Wishes*. The Duke of Marlborough is the only well-known military hero referred to in the poem and, as we shall see, he occupies a position of special significance within it.[5] It is not unlikely, moreover, that Johnson wished at the one and the same time to unmask his conquerors and to undo his unmasking, to reveal the limitations within the heroic and yet to preserve the heroic from the burlesque that reduces it by refusing to attribute any grandeur whatsoever to it.[6] The question asking those 'whose thoughts at humble peace repine/Shall Wolsey's wealth, with Wolsey's end by thine' (ll. 121–2) is balanced by the irony implicit in the injunction that closes the portrait of Archbishop Laud: 'Hear his death, ye blockheads, hear and sleep' (ll. 173–4). To ignore the latter injunction would be to turn anti-heroic irony into an instrument of mastery in the service of slaves and in so doing to revive the goal of self-sufficiency it had so strenuously denied.

This ambivalence is, of course, entirely consistent with the view that the warrior-hero, far from being autonomous, must rely upon the suppliant for his sense of self-worth. As Johnson puts it in *Idler*, No. 41, 'happiness is not found in self-contemplation; it is perceived only when it is reflected from another' (*Idler, Adventurer*, p. 130). In *The Vanity of Human Wishes*, the disparity between self-contemplation and reflection is manifested in the contradictory aspirations of the warrior-hero: a striving for self-sufficiency that is also craving for honour, fame, and popular applause. The impasse between these two irreconcilable desires is what provides the *raison d'être* for the inherent restlessness, the unappeaseable desire, that is so prominent a feature of the warrior-hero in the poem.

These two aspirations, moreover, are not equal in merit. The desire for self-sufficiency is conceived of as essentially heroic, while the craving for fame is viewed as servile and demeaning. Yet of the two impulses, the desire for fame is clearly predominant, as Johnson indicates in a general statement about epic heroism:

> The festal blazes, the triumphal show,
> The ravish'd standard, and the captive foe,
> The senate's thanks, the gazette's pompous tale,
> With force resistless o'er the brave prevail.
> Such bribes the rapid Greek o'er Asia whirl'd,
> For such the steady Romans shook the world;
> For such in distant lands the Britons shine,
> And stain with blood the Danube or the Rhine;
> This pow'r has praise, that virtue scarce can warm,
> Till fame supplies the universal charm.
>
> (ll. 175–84)

This dependence upon the 'universal charm' of 'fame' rather than the less potent appeal of 'virtue' is what enfeebles the conqueror's drive for power. Moreover, the source of this 'charm' in the applause of those over whom the warrior has triumphed in the universal struggle for dominion makes even its power suspect: the same logic that prompted Mohamet to question Irene's professed love prevents the martial hero from ever resting on his laurels. Failing to gain more than fleeting pleasure from the flattery of others, he becomes compelled to aspire to ever new heights of power, to acquire through conquest the glory he cannot attain through popular applause.

Glory, fame, and popular applause are, from this perspective, entirely illusory, and the threat posed by the warrior's unquenchable thirst for these evanescent satisfactions is that it will lead him to exceed his limits and thus ensure his downfall. To a certain extent, the characters of all

Johnson's warrior-heroes share in this pattern, albeit in different ways.[7] The pattern is most evident perhaps in the portrait of Xerxes:

> Great Xerxes comes to seize the certain prey,
> And starves exhausted regions in his way;
> Attendant Flatt'ry counts his myriads o'er,
> Till counted myriads sooth his pride no more;
> Fresh praise is try'd till madness fires his mind,
> The waves he lashes, and enchains the wind;
> New pow'rs are claim'd, new pow'rs are still bestow'd
> Till rude resistance lops the spreading god;
> The daring Greeks deride the martial show,
> And heap their vallies with the gaudy foe.

> (ll. 225–37)

Unable to find permanent satisfaction in a census of his conquered subjects, Xerxes falls prey to 'madness', the total obliteration of limits. Like the astronomer in *Rasselas*, he becomes transformed into a 'spreading god', a lord of himself who 'imagining himself' imagines himself subject to nothing external to himself and is thus prompted to launch his impious plan to span the Hellespont on his way to conquer Greece.

Charles Albert of Bavaria is filled with a similar 'hope of plunder and praise'. Yet because he is incapable of distinguishing 'Honour's flatt'ring bloom/Of hasty greatness' from a true assessment of his prospects in Austria, the 'bold Bavarian' experiences the bewilderment that, like madness, portends the downfall of Johnson's benighted heroes. Basking in false acclaim, Charles Albert rushes to his 'fatal doom' in a rash attempt to scale 'the dread summits of Cesarean pow'r' (l. 242).

The critical question about Xerxes and Charles Albert might be why Johnson introduces them into the poem at all. It is by no means evident that a critic who was subsequently to argue that biography is more useful to the common reader than history, private lives more delightful than public performances, should devote so much attention in *The Vanity of Human Wishes* to the fall from 'heigh degree' of heroic warriors. Unremitting emphasis upon the *libido dominendi*, a theme usually associated with history rather than biography, pervades the examples and unites the poem. For Johnson, however, this theme is universal in human nature and can manifest itself anywhere and at any time in human affairs. In the lives of public men, it can make itself felt as easily in a courtier like Cardinal Wolsey as in a warrior-monarch like Charles Albert. Indeed, so

similar is the pattern of Wolsey's life to those of Johnson's conquerors
that it is easy to overlook his links to Cali Bassa of *Irene* and to the
'Unnumbered suppliants/That crowd preferment's gate' later in the poem.
Like the martial heroes of *The Vanity of Human Wishes*, Wolsey cannot
help thinking of heroic aspiration in terms of honor's flattering bloom. Yet
because he basks in the reflected glory of Henry VIII, Wolsey cannot
appease his 'restless wishes' which depend for their gratification not on
his own authority but on the acquiescence of his 'suppliants' and
'followers' to the 'rays of regal bounty' which, through him 'shine' (l. 103).
Though his ambitions tower to ever 'new heights', they remain insatiable
because they encounter no real obstacles. This is the meaning of the
speaker's observation that 'conquest unresisted ceas'd to please,/And
rights submitted, left him none to seize' (ll. 107–8). Thus even before
Wolsey becomes a victim to the caprice of a tyrant whose favor he cannot
control, he experiences the dissatisfaction of a pseudo-conqueror whose
triumphs are hollow because they are attained over an enemy that has
already surrendered in advance.

It is significant that the reversal which brings about Wolsey's downfall
is embodied in the metonymic figure of the eye:

> At length his sov'reign frowns – the train of state
> Mark the keen glance, and watch the sign to hate.
> Where-e'er he turns he meets a stranger's eye,
> His suppliants scorn him, and his followers fly.

> (ll. 109–12)

The account of Wolsey's fall in this passage offers a striking contrast to
the description of him at the beginning of the *exemplum*. In that
description, Wolsey is shown governing in 'full-blown dignity' through
voice and gesture and through the reflected rays of another. Hence at that
point Wolsey appears, in a very real sense, to be blind, discovering his
own eyesight only later when he encounters the 'stranger's eye'. Though
it is not clear from the passage whether Wolsey marks the 'keen glance'
that leads to his downfall, it is evident that he apprehends the eye meeting
him at every turn as all-seeing. The eye is thus a gaze that circumscribes
him and transforms him into an object, a being who is perceived rather
than perceiving. It is not even necessary to suppose that Wolsey
experiences a real gaze. The 'stranger's eye' may be an imaginary eye, but
it is an eye which not only captures him by surprise but surprises him in
the disloyal posture of overweening ambition.

Wolsey is singular among Johnson's exempla in his readiness to govern with borrowed authority. Charles XIII, by contrast, aspires to become an all-conquering military hero. Convinced that he must subdue himself before he can subdue the world, Swedish Charles seeks to rise above the pleasures and pains to which ordinary mortals are subject:

> On what foundation stands the warrior's pride?
> How just his hopes let Swedish Charles decide;
> A frame of adamant, a soul of fire,
> No dangers fright him, and no labours tire;
> O'er love, o'er fear, extends his wide domain,
> Unconquer'd lord of pleasure and of pain;
> No joys to him pacific scepters yield;
> War sounds the trump, he rushes to the field;
> Behold surrounding kings their pow'r combine,
> And one capitulate and one resign;
> Peace courts his hand, but spreads her charms in vain;
> 'Think nothing gain'd, he cries, till nought remain,
> 'On Moscow's walls till Gothic standards fly,
> 'And all be mine beneath the polar sky.'

(ll. 191–204)

Yet in spite of Charles's efforts to transcend the contradiction between self-sufficiency and dependency, he comes to the same inglorious end as Johnson's other military figures. Insulating himself from the pleasures of popular approbation, Charles proves no more capable than Wolsey of appeasing his ambition. If Wolsey failed to derive satisfaction from the flattery of those who were forced to surrender to him, Swedish Charles fails to derive pleasure from his attempt to ignore the applause of those over whom he has conquered. Indeed, the more freely he avows his indifference to 'the charms' offered by 'peace', the more fully Swedish Charles becomes dependent on the need to undertake ever new conquests: 'War sounds the trump, he rushes to the field'. The asyndetic link between the halves of this line only serves to underscore the extent to which Swedish Charles, far from achieving autonomy, has not risen above a dependence to which he remains a prisoner. Swedish Charles becomes not 'the unconquered lord of pleasure and of pain', but the slave of a desire he has not created.

Thus the ironic reversal that divides the two halves of the portrait does not mark a genuine fall from heroic self-mastery to dependency, from conquest to supplication. From the very beginning, Charles's self-sufficiency was illusory. It is significant in this connection that Charles's military conquests and defeats are described throughout in theatrical terms. In fact, Charles is never without an audience. He is assured the constant attention of the 'nations' which 'on his eye suspended wait' and he does nothing to discourage them from paying attention to him. On the contrary, his call to arms seems designed to draw the world's gaze to his last and ultimate conquest:

> Think nothing gain'd, he cries, till naught remains,
> On Moscow's walls till Gothic standards fly,
> And all be mine beneath the polar sky.

(ll. 202–4)

So also is the inception of his 'march' on Moscow, which is described as beginning 'in military state'. The extreme difficulty he experiences in carrying out this march is an indication of the extent to which Charles has become blinded to the fact that for him the world has become little more than a stage on which he enacts his brilliant performances.[8]

Thus the theatricality of Charles's spectacular campaigns becomes one means by which servitude is shown to inhabit the very core of the *libido dominendi*. Through the inescapable presence of this histrionic element in human affairs, the potentiality for heroic action is created, but so also is a dependency, a need to perform before others that is typical of the slave, the buffoon, and the actor. This may be why there is virtually no equivalent in *The Vanity of Human Wishes* to the hierarchic taxonomies of sincerity and mimicry, surly virtue and obsequious vice that were so prominent in *London*. Johnson's presentation of the theatrical analogy seems different when viewed in this context: in *The Vanity of Human Wishes*, the hero and the parasite are one and the same. Indeed, virtually 'every state' reverts to the supplementary, exterior condition of dramatic representation and thus partakes of that lack of spontaneity and immediacy which, for Thales, were largely confined to the public displays of the 'cheated nation's' sycophants.

Yet if modern life is a stage and all its men and women merely players, who, then, is the spectator for their performances? Johnson finds the answer to the question in the legendary figure of Democritus. Implicitly denying a position of privileged detachment to the modern reader, he proposes the 'philosophic eye' of the ancient philosopher as an alternative to the submission of the mind to fame's 'universal charm'. To

Democritus's ironic gaze, 'Britain's modish tribe' appear as a company of players whose performances encompass 'the riches of pleasure' and 'the veils of woe'. The range of alternatives offered by this spectacle (comedy or tragedy) is itself a measure of the alienation by which a divided being seeks to recover itself through representation. But this is only the most preliminary of judgments: under the scrutiny of the philosophic eye, the comedy and tragedy of modern life are revealed to be only 'solemn toys' and 'empty shew'.[9] In employing these phrases, Johnson is undoubtedly referring to the sentimental comedy and declamatory tragedy that he described as usurping the realm of 'Nature' in the *Prologue Spoken by Mr Garrick at the Opening of the Theatre in Drury-Lane, 1747*. These were terms associated with the decline of the theater in eighteenth-century England and with the displacement of tragedy and comedy by 'farce', a genre that is defined in the poem by the reign of folly and the pursuit of illusory rather than real objects. In farce, actions ascribed by Aristotle to probability and necessity are superseded by actions in which effects are divorced from causes ('joys are causeless') and the passions to which characters are subject are portrayed as useless ('griefs are vain').

The political allusions encapsulated in *The Vanity of Human Wishes* are characterized by the motiveless and inconsequential actions enacted in farce. This is particularly evident in the paragraph in which the speaker invokes Democritus:

> Thou who couldst laugh where want enchain'd caprice,
> Toil crush'd conceit, and man was of a piece;
> Where wealth unlov'd without a mourner dy'd,
> And scarce a sycophant was fed by pride;
> Where ne'er was known the form of mock debate,
> Or seen a new-made mayor's unwieldy state;
> Where change of fav'rites made no change of laws,
> And senates heard before they judg'd a cause;
> How wouldst thou shake at Britain's modish tribe,
> Dart the quick taunt, and edge the piercing gibe?

> (ll. 53–62)

Here farce is apparent not only in the physical comedy of the mayor's 'unweildy state', but also in the abolition or reversal of cause and effect in the conduct of monarch and legislature. Even the narratives of Wolsey and Charles XII are farcical by this standard, for their 'falls' from power are transformed, through the suppression of causal explanation, into sudden, unforeseen collapses. The disruption of metonymic progression brought about by this suppression divides their lives into the farcical pattern of 'causeless joys' and 'vain griefs'.

The reduction of tragedy into farce is also the diminution of the patriot hero into the courtier who merges into the 'Unnumbered suppliants' that 'crowd preferment's gate'

> Athirst for wealth, and burning to be great;
> Delusive Fortune hears th'incessant call,
> They mount, they shine, evaporate, and fall,
> On ev'ry stage the foes of peace attend,
> Hate dogs their flight, and insult mocks their end.
>
> (ll. 73–8)

The humiliation to which actors were subject when they were hissed off the stage at the close of a bad eighteenth-century play becomes the measure of the inglorious end to which the courtier and 'sinking statesman' must submit. The occurrence of an overtly theatrical analogy at this point carries with it a corrosive irony, for the audience hissing the actors off the stage are themselves ensnared in the same folly of modern life. The 'fall' of the 'sinking statesman's portrait' is an apt emblem of this folly: our new opinion of the disparity between its 'form distorted' and the statesman's actual features corresponds to the disappearance of the 'morning worshiper', 'weekly scribbler', and 'dedicator' from his door (ll. 80–2). For reader and spectator, no less than for statesman, ours is inescapably a theatrical world of mediating representations. Nowhere is there unmediated self-contemplation or unreflected judgment.

What is farcical about the kind of fall exemplified by the sinking statesman is that it occurs in spite of the intentions of the personnage. The kind of irrational necessity that propels the action of farce leads the courtier into a conflict with the king regardless of his own capacity or motives:

> What gave great Villier's to th'assassin's knife,
> And fixed disease on Harley's closing life?
> What murder'd Wentworth, and what exil'd Hyde,
> By Kings protected and to Kings ally'd?
> What but their wish indulg'd in courts to shine,
> And pow'r to great to keep or to resign?
>
> (ll. 129–34)

The necessity to which these statesmen are subject is simply that of the historical situation which, though it takes into account 'their wish indulg'd in courts to shine', ignores their sentiments and conduct.

There can be little doubt that the necessity propelling Villiers, Harley, Wentworth, and Hyde to their dooms extends to the scholar as well as the courtier, the valetudinarian as well as the temperate man. Within the argument of the poem, Johnson shifts back and forth between heroic and unheroic states of mind: accounts of statesmen and martial heroes are balanced by portraits of the scholar, valetudinarian, and 'beauty'. In his famous portrait of the scholar (ll. 135–64), Johnson seems to reject the pastoral ideal, affirmed in both *Irene* and *London*, of existence lived on the margin of history, avoiding the entanglements and dangers of the 'passing world'. In the case of the valetudinarian and temperate man, the issue is posed in terms not of historical events but of the anguishing dilemma of age and disease. Though the 'virtues of the temperate prime' are contrasted to the vices of the valetudinarian who is described as a 'suppliant' to 'protracted life', neither is exempted from the malediction of the body. Stripped, by their creaturely natures, of their heroic aspirations, figures who have transcended the misfortunes of history become prey to the ravages of time:

> In life's last scene what prodigies surprise,
> Fears of the brave, and follies of the wise?
> From Marlb'rough's eyes the streams of dotage flow,
> And Swift expires a driv'ler and a show.

> (ll. 315–18)

The menace of age for the heroic mind thus manifests itself in the same kind of unmotivated reversals ('fears of the brave, and follies of the wise') that transform the 'causeless joys' of the warrior into 'vain griefs'. The fact that Marlborough and Swift are implicitly understood, as models of temperance, to have avoided the temptations besetting earlier instances of warrior and statesman is what makes Johnson's portrayal of 'life's last scene' so shocking. Coming as it does so close to the end of the poem it conforms to one of the conventions of farce – the brutal, unexpected ending that can be manifested comically in the fall through the trap door or tragically in the sudden appearance of agents of doom. The effect of the juxtaposition of the two figures, moreover, is metaphoric; with a precision brutal in its physical detail, Johnson finds these ancient adversaries united in the spectacle of their physical debacle and public humiliation.

In the light of this seemingly inexorable triumph of pain and suffering, it is not surprising that many readers have found the conclusion unconvincing, a pious coda that fails to overcome the conflicts arising from the universal struggle for emulation. Yet while this struggle is not overcome, it is displaced. As Bate has observed, the basic model for 'helpless man' in *The Vanity of Human Wishes* is the 'suppliant who crowds preferment's gate or prays for life protracted:

> Enquirer, cease, petitions yet remain,
> Which heav'n may hear, nor deem religion vain.
> Still raise for good the supplicating voice,
> But leave to heav'n the measure and the choice.

(ll. 349–52)[10]

What links this passage to themes enunciated earlier in the poem is the parallelism between preferment's gate and heaven's gate. The initial paradigm is not erased but simply inverted, and the term that was once negative is now positive. Moreover, the conclusion does not promise an easy and certain path to salvation. The 'eyes' which 'discern afar/The secret ambush of a specious pray'r' (ll. 353–4) are related metonymically not only to Democritus's 'philosophic eye' but also to the 'keen glance' that detects Wolsey's overweening ambition. It is true that the 'will resign'd' is an easier goal for 'helpless man' who, confronted by a heavenly rather than an earthly monarch, never finds himself in possession of 'a pow'r too great to keep or resign'. Yet 'the goods for man' which the 'laws of heav'n ordain' do not actually resolve the contradiction between self-sufficiency and dependency. Rather they provide the means by which 'celestial wisdom calms the mind/And makes the happiness she does not find' (ll. 367–8). Johnson is probably using the verb 'makes' here in the sense of 'supplies' or 'repairs' an existing lack or deficiency (*Dictionary*, to make, no. 57). As a supplement, celestial wisdom is exterior, outside the positivity to which it must be added, different from the happiness for which it is a substitute. Thus where natural happiness ought to be self-sufficient, celestial wisdom leaves the mind still committed to a life of yearning. To the heroic mind, it offers the risk of faith rather than the risk of death. To the unheroic mind 'panting for a happier seat', it offers death, 'kind Nature's signal of retreat' (l. 364), instead of pastoral retreat.

Johnson's religion is thus aimed at making man into an empty vessel which God alone can fill. And since in order to arrive at this conclusion, Johnson begins from a position quite close to that of Hegel, it is not implausible that we should recall the successive moments of the *Phenomenology*: lordship, bondage, stoicism, scepticism, and the unhappy consciousness. Of these moments, stoicism and scepticism cannot apply inasmuch as the 'petitioner' enters into a relation with a divine lord rather than adopting a posture of resigned self-sufficiency. What is omitted from this relation, nonetheless, is any sense that the wisdom which 'calms the mind' also offers the happiness that only 'heaven's sacred presence' can give. Lacking this presence, the petitioner is indeed a perfect picture of what Hegel called the unhappy consciousness.[11] His mind remains self-divided not because it is despairing or sceptical but because it has invested all plenitude in a heavenly master. The petitioner has been relieved of the anxieties of total despair, but he is still condemned to a life of yearning. Although celestial wisdom calms his mind, the petitioner, like Hegel's unhappy consciousness, needs this heavenly presence too much to achieve any kind of happiness in the here and now.

Thus there is no difficulty in explaining how this ending conforms to what we know of Johnson's later attitude toward death and particularly his own approaching death. Though wholly orthodox, even Augustinian in outlook, its Augustinianism is nonetheless incomplete; it acknowledges the reality of the petitioner's sinfulness but bars him from any direct access to the grace that is also a part of Augustine's mature religious vision.[12] Undeniably, there is a revelatory efficacy in the petitioner's encounter with celestial wisdom. This wisdom does actually calm the mind. Prayers do work. But the effect of this calming process is also to reassert the impossibility of the transcendence that was raised by the prospect of the divine presence. The referent of the petitioner's prayers turns out to be not divine plenitude but another supplement. Celestial wisdom is, in short, the prime paragon of human wishing. Its purpose is to confirm that the ultimate truth of celestial wisdom is still the truth of unlimited deferral.

5

'The Maze of Variation': Johnson's Philosophy of Language in the *Plan* and *Preface* to *The Dictionary*

Willard Van Orman Quine once held that the activity of the lexicographer 'is just as formal in spirit' as that of the 'grammarian'.[1] By itself, this observation cannot explain what, in Quine's own words, is 'formal' about the lexicographer's practice, but it is valuable in drawing attention to the systematic, theoretical dimension of his labours. This applies with particular cogency to the case of Samuel Johnson, since surprisingly little has been written about the theoretical assumptions behind his *Plan* and *Preface* to *The Dictionary*. There are of course aspects of Johnson's lexicography that have been scrutinized, including, most significantly, the traditional view that he was a linguistic authoritarian whose overriding aim was to fix the language.[2] We are now much more likely to see Johnson as a disillusioned realist who, though he felt a deep emotional attachment to the ideal of correctness, nonetheless came to recognize the impossibility of such an ideal and hence the necessity of subordinating it to the imperatives of custom and usage.[3] But to compress Johnson's ideas on language solely to this aspect is to fail to appreciate the subtlety of the theory of language embodied in the *Plan* and *Preface*. Only by approaching these texts from a double perspective, at once broadly semiotic (regarding signs in general and the place of Johnson's ideas about language within semiosis at large) and synthetic (seeking to reduce Johnson's ideas to a coherent system) can their complexity be grasped and with it the influence that they exerted as a central constitutive element in the great project they seek to justify and explain.[4]

* * *

The starting point of Johnson's theory of language can be found in his definition of the relation between words and things in the *Preface*:

Language is only the instrument of science, and words are but the signs
of ideas: I wish, however, that the instrument might be less apt to
decay, and that signs might be permanent, like the things which they
denote.   (*Works*, V, 27)

Johnson's statement points in the direction of what seems to be a clearly
articulated semiotics: like Aristotle and Locke, Johnson introduces a
conceptual third term (ideas) that mediates between words and things and
thus enables us to distinguish meaning from truth. In this passage, the
problem of the signification of words is separated, implicitly at least, from
the issue of reference. It becomes possible to study the words which make
up a language empirically and independently of what they 'denote' – as a
body of signs which are defined by their relation to other signs.

What seems to be only a conventional semiotic distinction has,
however, another, more disturbing aspect. This aspect can be seen in the
manner in which Johnson refers to the 'decay' of language. From this
viewpoint words seem to lack permanence, for, as Johnson puts it, they
are the daughters of earth', in contrast to things which *are the sons of heaven*'
(*Works*, V, 27). As '*the daughters of earth*', words are deprived of the
plenitude which would be necessary if they were to be invested with the
stability of their own self-identity. This lost plenitude, this divine absence
is reflected in the ineradicable exteriority which bars any strict
correspondence between words and things, language and science. As an
imperfect 'instrument', language is the contingent otherness of science
and thus of conceivable and signifiable meaning. As such, it is subject to
degeneration and decay. At the heart of Johnson's semiotics thus appears
to be a kind of rupture, a disjointing of the traditional antimony *res/verba*
that militates in favour of the solidity of things over the evanescence of
words.

It is tempting to suggest that the awareness of the inevitability of
linguistic change implied in this definition really links Johnson's outlook
with that of nineteenth- and twentieth-century theorists. On a second
look, however, we see that an awareness of change may refer to entirely
different perspectives. Linguistic change, as we have come to understand
it, is largely value-free and organic, probably associated with evolutionary taxonomies and completely divorced from any taint of corruption.
Change in Johnson's *Plan* and *Preface*, on the other hand, cannot be
deemed value-free, inasmuch as it is attributed to the fact that language is
'the work of man, of a being from whom permanence and stability cannot
be derived' (*Works*, V, 12). This remark, precisely because it is conceived
as germane to lexicography, expresses a dissatisfaction compatible with a
view of linguistic change as arbitrary, irregular, and even wanton. The

corruption here seems to consist of two things: the lack of any propriety, consistency, or normative authority in the production of linguistic change, and the independence of this kind of change from any attempts to arrest it. For a lexicographer like Johnson, habituated by temperament to preserving words and phrases, as far as possible, from decay, such an inevitability was bound to seem disconcerting. To try to fix the language becomes a way of encouraging the very change the lexicographer is trying to eliminate; to 'comply with the corruptions of oral utterance', however, is even worse, for it is 'to copy that which every variation of time or place makes different from itself, and imitate those changes which will again be changed, while imitation is employed in observing them' (*Works*, V, 27).

To the extent that this attitude toward language can be connected to Johnson's overall outlook — to the deeply and unselfconsciously held views that guided him throughout his career — it is perhaps best understood in relation to the postplenist perspective of his review of Jenyns's *Inquiry*. For the very notion of linguistic change, as Johnson envisions it, appears dependent on the notion of vacuity. In the conception of mutability embodied in the *Plan* and *Preface*, Johnson introduces vacuity, not only in the vertical chasm between words and things but in the horizontal maze of variation that constitutes the internal history of words. For Johnson, change does not imply continuity, the repetition of the same within the different, but introduces discontinuity, inaugurating the whole infinite series of the arbitrary and the contingent. A breach is opened in language, and inconsistency, limitless anomaly, rushes in and takes over. The nouns 'permanence' and 'stability' summon up all these conditions that are rendered impossible by this triumph: grammaticality, regularity, uniformity, analogy, rationality. In the absence of these conditions, language becomes subject to the infinite variations of what Johnson, throughout the *Plan* and *Preface*, calls human 'caprice'.

The commanding place that 'caprice' assumes with respect to historical change introduces a characteristic arbitrariness into the nature of the linguistic sign. For neither pronunciation nor orthography can exist by itself, and each is capable, by its faulty operation, of endangering the internal consistency of the other. Thus when 'oral' language:

> was first reduced to an alphabet, every penman endeavoured to express, as he could, the sounds which he was accustomed to pronounce or to receive, and vitiated in writing such words as were already vitiated in speech. The powers of the letters, when they were

applied to a new language, must have been vague and unsettled, and, therefore, different hands would exhibit the same sound by different combinations. (*Works*, V, 24–5)

The all-important consequence of this phonetic variation is that the idea of an alphabet naturally bound to the substance of phonic expression becomes impossible: 'from this arbitrary representation of sounds by letters proceeds that diversity of spelling, observable in the Saxon remains, and, I suppose, in the first books of every nation' (*Works*, V, 25).

That Johnson believes that the correlation of a given sign with its associated signifier or acoustic image is purely arbitrary was of course later made clear in his *Life of Pope*. There he launched his celebrated attack on Pope's view that the sound must seem an echo of the sense: although 'every language has some words framed to exhibit the noises which they express, as *thump, rattle, growl, hiss*, these words are but "few" and are thus purely marginal' (*Lives*, III, 230–1). Earlier, Johnson affirmed a modified version of this position in *Rambler*, No. 168, where he contended that while no word is 'intrinsically meaner' than another, it acquires positive or negative associations through 'custom' or 'accident'. In such circumstances, 'the mind' may indeed govern 'the ear, and the sounds' be 'estimated by their meanings' (*Lives*, III, 231). Yet associationism by itself cannot be said to naturalize the relations between sounds and their meanings. The fact that custom itself is mutable insures that the relation between meanings and sounds will invariably be contingent.

At the same time, the idea that linguistic change is inevitable makes it impossible to assume that speaking should be given priority over writing in the establishing of a language, so that the assertion that pronunciation should take precedence over etymology is only one possible hypothesis:

It has been demanded, on one hand, that men should write as they speak; but, as it has been shewn that this conformity never was attained in any language, and that it is not more easy to persuade men to agree exactly in speaking than in writing, it may be asked with equal propriety, why men do not rather speak as they write. (*Works*, V. 6–7)

Thus Johnson disposes of a question that had vexed critics of the past, namely the issue of linguistic norms. By refusing to presuppose a derivation of language from the substance of either phonic or graphic expression, Johnson places the problem outside the area of linguistics or lexicography. To the extent, of course, that Johnson believes that writing confers a permanence and stability absent in oral utterance, it can be conceived of as a necessary supplement to speaking in his lexicography. Yet this operation of supplementarity is not exhibited as a recuperation of the plenitude and rationality characteristic of general grammar: 'Dialects' will 'always be observed to grow fewer and less different, as books are multiplied', but 'the diversity of spelling, observable . . . in the first books of every nation . . . perplexes or destroys analogy, and produces anomalous formations, that being once incorporated, can never be afterward dismissed or reformed' (*Works*, V, 25).

Thus writing, which appears in all emerging nations, becomes increasingly capable of producing a measure of linguistic stability without, however, introducing a comparable degree of rationality. Johnson's emphasis on the 'lasting advantages' of this kind of stability combined with his hostility to linguistic innovation might seem to place him in league with Edmund Burke who similarly evoked tradition in later attacks on political innovation.[5] But in his opposition to linguistic reform, Johnson has in mind a conception of tradition quite different from Burke's vision of slow organic alteration and continuity. In Johnson's conception of tradition, the unstable conditions of oral utterance become an explanation of the reason why language must be arbitrarily arrested, if possible, at some point. This point does not possess any necessary identity, resemblance, or continuity with the language that preceded it. The field of language is organized by the principle of difference and discontinuity: the alteration of linguistic entities always reflects this principle, and no attempt to stabilize the language can ever erase it.

This conception of linguistic change has a much more profound effect upon Johnson's lexicographical programme than simply discouraging linguistic reform. It directs attention to the idea rather than the word. It is here than one finds the core of Johnson's project: in both the *Plan* and *Preface*, he assumes that change in orthography and pronunciation occur at a more rapid rate than analogous alterations in the ideas they signify. More susceptible to the caprice of the individual speaker or writer, the signifier has no natural attachment to the signified. Johnson's response to this discrepancy is to focus on meaning rather than material expression, to subordinate the sign as much as possible to thought. Orthography is to be settled wherever possible, and pronunciation is not to be 'wholly neglected', but the main emphasis of *The Dictionary* is to be placed on the 'signification' of words. Given the radical instability of the signifier, attention is inevitably directed to what Johnson calls the 'interpretation' or 'explanation' of the signified.

*   *   *

Yet if lexicography is not as susceptible to the same degree of variation as orthography and pronunciation, it nonetheless inextricably bound up with the words it seeks to define. In his recognition of this fact, Johnson resembles Francis Bacon who in *Magna Instauratio* had described the '*Idols of the Marketplace*' as 'the most troublesome of all idols' because they were concerned with 'the alliances of words and names'. Because the 'high and formal discussions of learned men' often end in fruitless 'disputes' over language, Bacon held that 'it would be more prudent' to begin with definitions and by definitions 'reduce' words 'to order'. Yet Bacon never believed that definitions would by themselves 'cure this evil', since definitions themselves 'consist of words, and these words beget others'.[6] Thus a part of Bacon's claim concerning the rationality of definitions rests on their tautological consistency – of their relation to things Bacon has little to say. Instead of pointing outwards to a supposedly represented *res*, definitions lead merely to other words and significations in an interpretive chain that produces nothing but a closed circuit.

Bacons's blending of rationality and scepticism strikingly anticipates the mood of Johnson's *Preface*. Although Johnson never claims that all definitions are tautological, his view that the task of interpreting 'a language by itself is very difficult' (*Works*, V, 34) leads one to look back at the sort of examples Bacon cites, where words are often 'loosely and confusedly defined', and to ask whether words that are 'irregularly derived from realities' do not depend upon accidental connections that put in question the truth-value of the claims made for definitions of such words. Or to put it somewhat differently, if 'nothing can be defined but by the use of words too plain to admit a definition' (*Works*, V, 34), won't a certain circularity be a necessary feature of the meanings disclosed?

This fundamental question about the nature of definitions does not deter Johnson from undertaking the task of interpretation. But it does lead him to raise issues about the difficulties involved in this task. For Johnson, these difficulties are the logical consequence of the three-fold nature of the linguistic sign and thus can be related to things, ideas, and words. Of the first kind of difficulty are signs which cannot properly be said to have referents because 'the nature of things is unknown'. Although this group of signs is cut off from any standard of truth, it is not a major source of concern for Johnson as a lexicographer, since he believes that 'some obscurities' can be left 'without shame' to 'happier industry, or future information' (*Works*, V, 35). The elucidation of such 'obscurities' thus depends on a certain type of factual or regional information (historical, botanical, chemical, etc.) giving rise to a knowledge that is itself regional and, as such, at least partly beyond the domain of the lexicographer who deals with the meaning of known words.

But the second class which stems from faulty usage is more complex and deeply ingrained in the limitations of the human psyche. Like Bacon, Johnson argues that this group must be studied within the field of human consciousness. What this means is that the practical difference between precise and imprecise usage is parallel to the philosophical difference between distinct and indistinct ideas. Thus a word may be indeterminate – that is, it may designate more than one meaning, or it may be confused and ill-defined and hence allow one to speak, as Johnson does, of 'notions unsettled and indefinite and various in various minds' or of 'significations . . . so loose and general, the use so vague and indeterminate' that it becomes difficult

> to trace them through the maze of variation, to catch them on the brink of utter inanity, to circumscribe them by any limitations, or interpret them by any words of distinct and settled meaning. (*Works*, V, 35)

Such difficulties stem from the fact that definitions must be grounded in common usage. Definitions that record the different meanings of a word are reports on popular, not philosophical usage. But such reports can become part of a language only insofar as usages are clear and distinct, each one remaining singular and identifiable. Definitions are definitions only insofar as they can recover and define these usages; when they try to record vague and unsettled notions, they become damaged or even worthless. There is, however, one kind of definitional activity that seemed to promise a way out of this morass of linguistic imprecision. Johnson, like other lexicographers of the eighteenth century, believes that 'every word of common use' has but one dominant, proper, literal, natural sense. This dominant sense, by extension, is also regarded as its primitive as opposed to secondary, derivative senses. Johnson believes, or professes to believe, that the *etymon* of this primitive sense always exists, however hidden it may seem. Such a primitive sense is intelligible in and of itself and has an immediate relation to an object or rather to a clear and determinate unity of meaning. It constitutes a normative order of signification from which all 'remote or accidental' significations are excluded as corrupt and derivative. Ideally, therefore, the lexicographer must be an etymologist, he must 'mark' the 'progress' of every common word and 'show by what gradations of intermediate sense it has passed from its primitive to its remote and accidental signification; so that every foregoing explanation should tend to that which follows, and the series be regularly concatenated from the first notion to the last' (*Works*, V, 36).

The recurring praise of this linear model found in Johnson's *Plan* and *Preface* should not blind one, however, to his doubts about its validity. In giving voice to such doubts, Johnson clearly differs from a linguist like Anselm Bayly who in his *Introduction to Languages* (1738) argued that if we trace words back to their primitive senses, 'it will lead us through most of the mazes of language, and open to us such a view of things, such a perspicuity and precision in our ideas, as otherwise are unthinkable'.[7] By contrast, Johnson emphatically avoids such expressions of methodological euphoria. Thus, while he professes to subscribe to the etymological ideal embodied in Bayly's statement, he maintains that in practice, it frequently proves impossible to reactivate the primitive inscription and restore the palimpsest:

> kindred senses may be so interwoven that the perplexity cannot be disentangled, nor any reason assigned why one should be ranged before the other. When the radical idea branches out into parallel ramifications, how can a consecutive series be formed of senses in their nature collateral? The shades of meaning sometimes pass imperceptibly into each other, so that though on one side they apparently differ, yet it is impossible to mark the point of contact. Ideas of the same race, though not exactly alike, are sometimes so little different, that no words can express the dissimilitude, though the mind easily perceives it, when they are exhibited together; and sometimes there is such a confusion of acceptations, that discernment is wearied and distinction puzzled, and perseverance herself hurries to an end by crowding together what she cannot separate.   (*Works*, V, 36)

At this point in Johnson's argument, the relation between the derivative senses – which are described as taking off in unforeseen and unpredictable directions – and the primary sense seems quite tenuous. For in dissolving boundaries and limits, secondary meanings break the tie that binds them to any primary signification. The history of a word appears not as a linear series in which the meaning of a term is clearly and distinctly marked off from those which precede or follow it, but rather as a genealogical tree in which senses might be parallel or even as a spectrum in which senses might differ yet seem to merge insensibly into one another.

Johnson thus examines the problem of etymology, on which genetic theories of signification in the eighteenth century were frequently based, and gives it an emphasis which seems to cast doubt on the value of etymological speculation. But the opposition between primary and secondary meanings was not the only source of such speculation. Behind that opposition lies another, even more fundamental contrast – the contrast between literal and metaphoric significations. The etymologist's primitive meaning is a literal meaning, but this meaning can readily be displaced or even effaced by metaphoric derivatives: 'the original sense of words is often driven out of use by their metaphorical acceptations, yet must be inserted for the sake of a regular origination' (*Works*, V, 37). In the *Plan*, Johnson, much more confident that the *Preface* of his ability as a lexicographer to trace the chain of significations back to their source, provides a genetic trajectory of metaphoric signification. This trajectory moves from the 'natural and primitive' through 'metaphoric' and 'poetic' senses, to 'familiar', 'burlesque', and 'peculiar' senses (*Works*, V, 15). To the extent that this pattern can be taken seriously, it projects a history of the word, not as a displacement with breaks, as reinscriptions in a heterogeneous and indiscernible maze, but rather as a serio-comic rise and fall, an ascent to a height of elegance and grace, followed by a plunge into the depths of the burlesque and the common. In the background of this trajectory is an implicit history of English poetry from Milton through Pope to Gray – a history that traces a movement from rudeness to refinement and thence to corruption.

In the *Preface*, as we might expect, Johnson tacitly abandons this schema, preferring to emphasize the accidental and capricious rather than the necessary and genetically determined in etymological speculation. But the confusions engendered by the temporal concatenation that constitutes etymology was by no means the only source of difficulty confronting the lexicographer. Johnson stands at the end of a long tradition in which nouns and verbs are clearly marked off from other parts of speech. What nouns and verbs have in common is that they point to substances and actions and thus stand in oppositions to the articles, conjunctions, and prepositions which, according to this tradition, have no meaning in themselves and which do not of themselves denote anything. Thus even if nouns and verbs are recognized to be subject to the erosion of time, the other elements of speech are marked from the outset by a 'latitude' (i.e. spatial variation) that renders even their incorporation within a 'regular scheme of explication' marginal. These elements of speech include 'particles' and 'expletives', which, in dead languages, 'are suffered to pass for empty sounds, of no other use than to fill a verse, or to modulate a period, but which are easily perceived in living tongues to have power and emphasis, though it be sometimes such as no other form of expression can convey' (*Works*, V, 34–5).

There is thus no part of speech which is not subject to lexicographical doubt. If nouns and verbs are privileged over the other parts of speech, they are subject to a temporal erosion that makes them subject to the same distortions and errors as prepositions and adverbs. The only way to minimize the possibility of error is to view words, not merely in terms of meaning and reference, but also as relational entities, as products of a system of differences. This is in effect what Johnson proposes to do in the *Plan*. To define a large number of words, it is necessary, he contends, to explain them 'by their opposition to others, for contraries are best seen when they stand together'. The meaning of the verb 'to *stand*', for example, is not merely a representation in one's mind but a space in a complex network of differences in which it can be opposed, on the one hand, to the verb 'to *fall*' and, on the other, to the verb 'to *fly*'. Johnson uses this example as the focal point of a series of oppositions: 'strict and c̈ritical' vs. 'loose and popular', 'active' vs. 'passive', 'personal' vs. 'real'. In this way, Johnson seeks to take up the problem of definition and give it a relational and differential interpretation which seems not only to promise a new type of explanation – a structural explanation in terms of an underlying system of relations – but also to displace the ponderous and futile etymological labours of earlier scholars like Richard Bentley (*Works*, V, 16–17).

By the time he came to compose the *Preface*, however, Johnson also seems to have dismantled this structuralist system of hierarchic relations.[8] In a shift reminiscent of the transition from *London* to *The Vanity of Human Wishes*, Johnson finds these oppositions collapsing and discovers in the process a third kind of lexicographical difficulty – one that relates not to things and ideas but to words themselves. Against his earlier antithesis of the critical to the popular, Johnson now observes that 'to explain, requires the use of terms less abstruse than that which is to be explained, and such terms cannot always be found' (*Works*, V, 35). Thus rather than setting one off against another, Johnson now recognizes that the critical term requires as its supposition the very popular term it initially meant to oppose. Moreover, the translation from critical into popular has a limit, for 'the easiest word, whatever it be, can never be translated into one more easy' and, indeed, 'easiness and difficulty are merely relative' (*Works*, V, 38). Johnson goes a step further, questioning definitions based on synonymy as well as antonymy: 'many words cannot be explained by synonyms, because the idea signified by them has not more than one appellation; nor by paraphrase, because simple ideas cannot be described' (*Works*, V, 34). Just as English philosophers from Locke to Hume assumed that all things are individual, so Johnson avers that 'words are seldom exactly synonymous: a new term was not introduced, but because the former was thought inadequate: names, therefore, have often many ideas, but few ideas have many names' (*Works*, V, 36). In the absence of such synonyms, the lexicographer is forced to rely on 'the proximate word, for the deficiency can seldom be supplied by circumlocution' (*Works*, V, 36).

The circumlocution which is essential to the overall success of the definition is, Johnson implies, often 'unavoidably reciprocal or circular' (*Works*, V, 37–8). Such a meaning cannot be interpreted. It cannot interpret itself, or be interpreted by what it has generated. It can only be explained by something outside of itself, outside the system of relations that constitutes the verbal definition. Johnson finds this something in 'the examples, subjoined to the various senses of each word'. It is only here that one can find 'the solution of all difficulties, and supply of all defects' (*Works*, V, 38).

It might be objected that Johnson's cavils about the difficulties inherent in the 'interpretation' of words in the *Preface* are largely tactical – that is, they are intended to disarm criticism and to highlight the actual achievement of *The Dictionary*.[9] But while there is undoubtedly some truth to this argument, it does not tell the whole story. As we have seen, the *Preface* abandons two schema that were central to Johnson's programme in the *Plan*. The first was diachronic – a paradigmatic history of the word that charted a trajectory of progress and decline; the other was synchronic – a series of oppositions by which the definitions of words were to be integrated into a system of differential relations. In implicitly recognizing the limitations of both schema, Johnson was probably reverting to a Renaissance scepticism about language rather than looking forward to a nineteenth-century historicism. Abandoning the confident linguistic rationalism of the late seventeenth and early eighteenth centuries, Johnson came to acknowledge a problematic that can be found in the writings of Bacon and his other favourite Elizabethan authors. Indeed, it is the very depth of Johnson's commitment to this problematic that informs his theory of language in the *Preface* – so much so that it is worth reflecting at greater length on the underlying assumptions of this theory.

* * *

The characteristic features of Johnson's linguistic scepticism can be better understood when they are placed in the context of eighteenth-century speculations concerning the origins of language. As Hans Aarsleff has noted, the eighteenth century was preoccupied with the search for origins, whether of institutions, society, government, the family, or language. Aarsleff observes that this notion of origins was not historical and factual in the nineteenth-century sense of the term. Rather it was as hypothetical, he argues, as 'the state of nature in political philosophy and like the latter, its aim was to understand men in the present'.[10] It is possible to distinguish two kinds of speculative origin in eighteenth-century linguistic thought: a neo-Platonic divine institution which might also call to some extent upon *Genesis* and might allude to Adam, Cain, Noah, and the Tower of Babel; and a natural origin which might speculate upon a pre-linguistic state in which a language of gestures and inarticulate cries linked men to animals. Though natural explanation undoubtedly took precedence over theological explanation in eighteenth-century thought, both project a point of presence, a plenitude which assures the stability and coherence of linguistic structures.

The degree to which Johnson was concerned with the origins of language reflects, therefore, a tendency typical of his age. Like many of his contemporaries, Johnson argues in the *Plan* 'that speech was not formed by an analogy from heaven. It did not descend to us in a state of uniformity and perfection' (*Works*, V, 11). This denial of divine origins, needless to say, is also a denial of presence and plenitude. Divine action provides the impetus for the model of a perfect language, a model which certainly fascinated Johnson in the *Plan* and *Preface*. In both texts, the principle of analogy in the verbal world corresponds to the principle of continuity in the universe. If the different orders of creation are linked by an infinite series of gradations, then the different parts of language are also articulated by analogy into an organic whole.

But if Johnson resembles many of his contemporaries in rejecting this theologically-based picture of language, he differs from them in refusing to consider an alternative natural source. As John Barrell has noted, Johnson avoided 'the theoretical questions concerning the origin of language and government that preoccupied so many political thinkers from Hobbes and Locke and throughout the eighteenth century'.[11] In the eighteenth century, the atomic hypothesis offered an answer to one such theoretical question. In place of a universe governed by divine will or an ultimate end, it held that particles of matter, though directed by chance, are nonetheless preserved throughout time from annihilation. These particles exist as plenitude, not because they are guided by a divine *telos*, but because they are primordial and hence indestructible. Johnson, however, refuses to attribute even this kind of permanence to speech whose elements, he insists, lack the duration we gratefully attribute to things:

> Who . . . can forbear to wish, that these fundamental atoms of our
> speech might obtain the firmness and immutability of the primogenial
> and constituent particles of matter, that they might retain their
> substance, while they alter their appearance, and be varied and
> compounded, yet not destroyed.   (*Works*, V, 12)

Theoretical explanations of the origin of language – whether
theological or naturalistic – do not, however, completely account for the
eighteenth-century infatuation with genetic theories of language.
Another, equally important explanation was the supposedly historical
origin that Johnson sometimes ascribes to the English language. In one
passage from the *Preface*, for example, he argues that:

> our language, for almost a century, has, by the concurrence of many
> causes, been gradually departing from its original Teutonick character,
> and deviating towards a Gallick structure and phraseology, from which
> it ought to be our endeavour to recall it by making our ancient volumes
> the groundwork of our style, admitting among the additions of later
> times only such as may supply real deficiencies, such as are readily
> adopted by the genius of our tongue, and incorporate easily with our
> native idioms.   (*Works*, V, 39–40)

At such points as this, the argument is developed through a system of
oppositions, pure vs. corrupt, primary vs. secondary, original vs.
derivative that recall the country ideology of *London*. These oppositions
participate in a metaphysic of origins that Johnson appears to reject
whenever he discusses the theoretical beginnings of language. At other
points, however, Johnson employs a vocabulary that seems to call even
this mystique of origins into question. Thus when discussing the
derivatives of words in the *Plan*, Johnson contends that 'our language is
well-known not to be primitive or self-originated, but to have adopted
words of every generation, and, either for the supply of its necessities, or
the increase of its copiousness, to have received additions from very
distant regions' (*Works*, V, 10). In such passages as this one, Johnson
makes it clear that he believes the English language contains from the
very outset all the secondary and derivative attributes that he ascribes
elsewhere to its Gallic corruption.[12]
Yet if Johnson's repudiation of the myth of historical origins is not
nearly as decisive as his repudiation of the fiction of theoretical origins, it
is nonetheless significant enough for us to conclude that Johnson refuses
to find in origins any sustained source of coherence and linguistic

identity. Describing his initial 'survey' of the English language, Johnson reports that he found it 'copious without order, and energetick without rules: wherever I turned my view, there was perplexity to be disentangled, and confusion to be regulated; choice was to be made out of boundless variety, without any established principle of selection' (*Works*, V, 23–4).

What are the implications of this state of 'boundless' linguistic 'chaos' for the English language? It seems clear that Johnson rejects the assumption that the living language ever existed as an institution, or what Ferdinand de Saussure called *langue*, prior to and independent from its expression in speech acts, or what Saussure termed *parole*. For Johnson, the language that emerged from these speech acts was fragmentary, its contours blurred, its parts anomalous, and syntax imprecise. This does not mean, however, that Johnson subscribes to the theory, common in his age, of the existence of a primitive English speech without parts and grammar, a natural language of pure effusion. Such a language would only be another form of plentitude, a product of natural necessity that could provide the basis for a teleology of organic growth and development. By way of contrast, Johnson holds that speech is the product of both 'necessity' and 'accident' and is thus composed of 'dissimilar parts' which do not exist on a 'state of uniformity' but which are rather 'thrown together by negligence, by affectation, by learning or by ignorance' (*Works*, V, 11). Described in this way, speech resembles not so much Saussure's *langue* as his *langage*. The latter, Saussure holds, 'is many-sided and heterogeneous'.[13] Rather than embodying a deep structure or a coherent system of differential relations, speech is composed of a *bricolage* of elements which, though coherent in certain parts, nonetheless, does not present the 'uniformity' of a divinely or naturally instituted language. The linguist has to make do with this *bricolage*, however, because he has nothing else at his disposal. But his acceptance of this situation gives rise to no real solution, for it cannot lead to a coherent grammar or syntax.

Thus what distinguishes the 'dissimilar parts' of Johnson's speech from parts which might have been instituted on the basis of 'an analogy sent from heaven' is that they cannot be subordinated to a coherent system of classification, a universal grammar. In the absence of a grounding centre and origin, 'inflections . . . are by no means constant, but admit of numberless irregularites', and these 'irregularities' remain stubbornly 'anomalous' — that is to say, that they remain inscribed within a heterogeneous mixture of resemblances and differences which 'cannot be reduced to rules' (*Works*, V, 11–12). Unlike the stable, synchronic system of the Saussurian *langue*, this mixture is never conceived as resting for a moment in the stability of its own existence. It must continuously repeat itself in a movement that is bound to remain endless. This movement persists regardless of any illusion of intentionality, regardless of the way in which its elements might be reconstituted into an internally coherent grammar. In the absence of a stable centre and origin, the resulting pattern of repetition and variation opens up and extends the play of signification infinitely.

As the transformation of a heterogeneous mixture into a uniform *langue* cannot logically be produced from within an unstable system, one is led to conceive of an external agency. This apparently 'arbitrary' solution corresponds to the limitation of human speech and thus resolves many exigencies. In the *Plan* Johnson contemplates on the possibility that the lexicographer might serve as such an external agent, a kind of linguistic 'legislator' who, by fiat, will impose an order and stability on the chaos of living speech. Yet Johnson's gestures in this direction are certainly half-hearted and, as commentators have recently suggested, may have been dictated by Johnson's wish to cater to the purist expectations of his patron, the earl of Chesterfield.[14] In any event, these gestures are cancelled by Johnson's implicit recognition of the impossibility of such an enterprise. Words are condemned to exist as evanescent entities. They are always susceptible to the possibility of alteration, never to the promise of permanence.

Johnson thus takes up the problem of the origins of language, on which eighteenth-century notions of linguistic plenitude were based, and gives it an emphasis that seems to call into question, at least as far as English speech is concerned, the existence of a centered structure. One outgrowth of this view of the English language may very well be Johnson's interest in lexicography rather than grammar or syntax.[15] Johnson's conception of his native tongue seems to conform to Saussure's observation that 'languages in which there is least motivation are more *lexicographical*; and those in which it is greatest are more *grammatical*'.[16] The pertinence of Saussure's observation for Johnson is illustrated in Johnson's contention that 'the syntax of the [English] language is too inconstant to be reduced to rules and can only be learned by the distinct consideration of particular words as they are used by the best authors'. It is this empirical contingency which requires the suspension of the doctrine of analogy, the foundation of universal grammar. 'Our syntax . . . is not to be taught by general rules', Johnson argues, 'but by special precedents'. In this way, linguistics dissolves into lexicography, conceived not in terms of a Roman but an Anglo-Saxon legal analogy. Instead of trying to legislate by 'general rules', the lexicographer searches for 'special precedents', that is, for illustrations which can serve as examples and justifications for the definitions he gives. These definitions represent the particular meanings which, in the absence of a general grammar, must serve as the matrix for the study of language.

This is a coherent and comprehensible programme: since words acquire meaning in specific contexts, we must discover how they are actually used if we wish to understand them. But the individuality, the unrepeatability of words as they are used in specific quotations, reveals a danger inherent in Johnson's program and in the philosophical orientation of which it is the culmination. The strength of Johnson's analysis lies in the fact the he clearly recognizes this danger. To avoid finding a word whose meaning is not 'apparently determined by the tract and tenor of the sentence' (*Works*, V, 40), he subjects it to an analysis which isolates and enumerates its distinct senses. The word is thus seen not merely as the embodiment of a specific usage, but as the manifestation of a pre-existent meaning. Yet Johnson, like later theorists of language, recognizes that the value of a word, the value of our understanding of the word, cannot be contained in that pre-existent meaning but must be comprehended in its precise, innovatory usage. To establish that 'usage', Johnson supplies 'special precedents' drawn from the best authors, precedents that whenever possible 'gave a definition of a term, or such an explanation as is equivalent to a definition' (*Works*, V, 40).

The second perspective acts as a critique of the first; it seems to bring about a reversal, explaining meanings not by simpler words but by 'special precedents'. However, the first perspective also acts as a critique of the second in turn, for examples of usage are themselves made comprehensible by the meaning of the words which constitute them. The semiotics of lexicography thus gives rise to a movement in which each pole of an opposition (definition vs. particular usage) can be used to show that the other is in error, but in which the undecidable dialectic gives rise to no synthesis because the antimony is inherent in the very nature of language itself.

This antimony is but one variation of a general situation in which *langue* is opposed to *parole* in Johnson's theory of language. Unlike Saussure, Johnson does not assume that individual speech acts are necessarily a material manifestation of a *langue* conceived of as a self-contained structure, and that such outward manifestations (which vary from case to case) are irrelevant to the study of language as a system of signs. He rather believes that language belongs to both the individual and society and cannot be isolated from individual speech acts as a self-contained whole and principle of classification. The reasons for this belief are threefold. In the first place, Johnson differs from Saussure in assuming that linguistic phenomena experience an extremely rapid, not slow, rate of historical change. Altering very quickly and, with few exceptions, 'by the caprice of every tongue that speaks the language', they are in that respect similar to economic, political, or religious structures whose rates of change can, at certain moments, be quite rapid. Second, this characteristic instability is sustained by the unusual captiousness of the linguistic subject within it: behind the mobility of the langue lies the volatility of the individual speech act, a volatility alluded to in Johnson's metonymic reference to the 'tongue'. For Johnson, *parole* is a material and therefore mutable manifestation of the ever-present temptation to phonocentric illusion and is thus irreducible to a *langue*. Indeed, the notion of voice as presence is subject to no natural constraint whatsoever: words 'are but wind' and, like the vowels that were thought to be formed by the breath that comes from the lungs, can readily be compared to a meteorological disturbance. Johnson refers specifically to a class of verbs whose 'signification is so loose and general' that their meanings 'can no more be ascertained in a dictionary, than a grove, in the agitation of a storm, can be accurately delineated from its picture in the water' (*Works*, V, 35).

Finally, Johnson's implicit prescription for linguistic regeneration in the *Plan* and *Preface* assumes that degeneration is caused by individual acts of speaking and writing. He believed that *parole*, though volatile and not easily reducible to a system, nonetheless exerted a considerable impact on language. Even minor authors, whose writings have passed into oblivion, 'differ in their care and skill' (*Works*, V, 25) and thus possess the capacity to alter the language in unfortunate ways. This is why Johnson decides to choose what he believes are commonly regarded as the best authors for his illustrations. Convinced that language is subject to a multiplicity of influences, Johnson assumes that is is incumbent upon the lexicographer to choose examples from those who are revered as 'masters of elegance and style'.

But these examples serve only as limiting constraints on meanings they are intended to illustrate. There is, however, another difficulty in which the volatility of *parole* can be discerned. This is the permanent 'superabundance' of the signified under the signifier: 'names . . . have often many ideas, but few ideas have many names' (*Works*, V, 36). For Johnson, this basic asymmetry is paradigmatic of the opacity and ambiguity of language: 'most men think indistinctly, and . . . cannot speak with exactness; and, consequently, some examples might be put indifferently to either' of two closely allied significations (*Works*, V, 42 –

3). This sliding of the signified under the signifier poses a potential danger to the definitional activity itself. Because of the dearth of words, it becomes difficult, if not impossible, for the lexicographer to establish a stable system of correspondences between words and meanings. Indeed, the proliferation of meanings, though co-extensive with the conscious mind, perpetually threatens to undermine the illusory integrity of the words they are supposed to explicate.

The attack on the stability of the sign inherent in the notion of an asymmetry between names and ideas has a predictable consequence on the status of the lexicographer as a legitimating author. Eschewing any pretense of objectivity or certitude, Johnson makes his presence as subject felt throughout the *Preface* and *Dictionary*. But Johnson does not use the presence as the basis for asserting his authority to stabilize the language. Indeed, the most moving passages of the *Preface* are those in which Johnson acknowledges his own fallibility as a maker of dictionaries. The key to those passages is not to be found in Johnson's awareness of his greater or lesser control over the cognitive resources of the language but in his acknowledgment that an instability inherent in the language itself is what renders totalization impossible.

The *Preface*, the great testament to Johnson's aspirations as a lexicographer, bears witness to this sense of failure. Yet it is important to emphasize that Johnson's belief in the limitations inherent in these aspirations never gave rise to linguistic nihilism. Jacques Derrida once argued that 'totalization can be judged impossible' in either one of two ways: 'from the standpoint of the concept of *play*' or 'from the standpoint of a concept of finitude as a relegation to the empirical'. Of these two alternatives, Johnson chooses the latter. Thus he defines lexicography, in Derrida's terms, as 'the empirical endeavour' of a 'subject' confronting 'a finite richness' he can 'never master'.[17] The source of this richness is the absence of a grounding centre or origin which, according to Derrida, can be measured from the standpoint of play. But one might argue that this absence can just as easily be viewed from the standpoint of empirical investigation. Though such an absence can be recognized and analysed, discussed in terms of a theory, it can never be used as an excuse for not at least trying to exhaust the potentiality of the finite, of the empirical.

6

The Choice of Life: Art and Nature in *Rasselas*

The relation between *Rasselas* and the post-Newtonian perspective of his review of Jenyns's *Inquiry* can be seen in a conception of desire which, inasmuch as it is located on a horizon of vacuity, can never be realized in the natural world. The movement of desire is that of a poignant yearning and a disillusioned satiety: the deficient 'tedium' of an expectation that precedes 'endeavour' and the excessive 'disgust' of a 'fulness' that follows attainment. Given this play of defect and excess, it is not surprising that the central issue for the characters of *Rasselas* becomes the means of achieving lasting satisfaction or happiness. In Johnson's Oriental tale, the attempt to resolve this issue often takes the form of a distinctive conception of art in which the natural world is subjected to a logic of choice, and reorganized around a contrast between good and evil, purity and impurity, inside and outside. Within this system, art serves as the *techné* by which the 'blessings' of nature are systematically 'collected' and 'its evils extracted and excluded'.

In *Rasselas*, this paradigm is pursued as a guiding thread not only in the Happy Valley, but also through the successive variants of the Happy Valley that present themselves to the tale's protagonists, Rasselas and Nekayah, as they embark on their quest for happiness. Yet in each case what is noteworthy is the way in which art as the supposedly privileged term of a hierarchical opposition is subverted by what it has displaced. Because nature serves as the negative term of this process of subversion, it functions as a more active, anarchic, impure mode of existence. Yet because the reversal of the relations between art and nature can lead in turn to a new hierarchy based on the superiority of variety over uniformity, activity over passivity, outside over inside, it is susceptible to the same process of subversion as the term it deposed. The movement between them is thus less a consistent 'deflation' of one term by another as an endless series of reversals in which one, alternately, takes priority over the other.[1]

*　　*　　*

The conception of art that shapes life in the Happy Valley is established in the very first chapter of *Rasselas*. It is embodied in the system of classification that divides animal species into two categories governed by the polarity of inside to outside:

> All animals that bite the grass, or browse the shrub, whether wild or tame, wandered in this extensive circuit, secured from beasts of prey by the mountains which confined them. On one part were flocks and herds feeding in the pastures, on another all the beasts of the chase frisking in the lawns: the sprightly kid was bounding on the rocks, the subtle monkey frolicking in the trees, and the solemn elephant reposing in the shade. All the diversities of the world were brought together, the blessings of nature were collected, and its evils extracted and excluded. (1, 38)

Within the basic division outlined here, there is a further subdivision of animals that bite the grass or browse the shrub into 'wild' or 'tame'. This division, in turn, yields another degree of specification – a distinction between ' flocks' (sheep), 'herds' (oxen), and 'beasts of the chase'. Johnson, following the nomenclature of John Cowell in *The Interpreter*, defines the last group in *The Dictionary* as 'the buck, the doe, the fox, the marten, and the roe'. The case of the kid, the monkey, and the elephant is somewhat different, since the references to habitat imply a vertical distinction between high ('rocks'), middle ('trees') and low ('shade').[2]

In the creation of the Happy Valley, the principle of dichotomous division demands a third term – the mountains – which provide a boundary line between inside and outside. The origin of division and difference, this boundary line enables the Happy Valley to gather its blessings, to enclose itself around the security of its seemingly impassible obstacles by appearing to expel or exclude all that threatens it. The act of exclusion thus constitutes the otherness of the evil that endangers the Happy Valley by threatening to invade it from the outside, and the apparent symmetry of the opposition between good and evil contributes to the illusion of a closed system.

Yet this illusion is obviously dispelled not only by the desire of Rasselas, Nekayah, and their guide, Imlac, to escape from the Happy Valley, but also by their belated discovery of the boredom and misery of all the inhabitants. Johnson is at pains to imply that this was all along the underlying assumption and not just a disclosure introduced with the wisdom of hindsight in order to justify the discontent of Rasselas and Nekayah. Imlac points out that the restrictions imposed on 'liberty' in the

Happy Valley, far from contributing to 'a community of love or esteem', actually appear to provoke the natural passions of malice and envy that they were meant to banish (12, 70). These passions are the psychological concomitant of an artificial system which, however tolerant it may appear, is as repressive as the natural anarchy it has presumably replaced.

This reversal of expectation carries a particular significance, for it corresponds in form to the founding myth of the Happy Valley. In this myth, the palace, which was erected 'as if suspicion herself had dictated the plan', could actually banish strife only by granting a special place within its walls to the heirs of the throne. These heirs were of course imprisoned by the ruler who lodged them at his own expense, but their presence within the precincts of the palace is emblematic of the inverted logic by which an element supposedly excluded from the Happy Valley is shown to inhabit its privileged space. That this myth figures in the narrative largely by indirect allusion – that it belongs, so to speak, to the sources of *Rasselas* – cannot prevent it from obtruding metaphorically into the progress of the action. Nowhere is its presence more pronounced than in the restrictions promulgated to make sure that 'impotence precludes malice' and that 'all envy is repressed by community of enjoyments' (12, 70). What the narrative brings out most strikingly is the extent to which these restrictions were formulated, all along, from the point of view of the 'regions of calamity' they were supposedly excluding.

It is precisely because of this perverse logic that the questions troubling Rasselas, Nekayah, and Imlac do not disappear when they leave the Happy Valley.[3] The binary, reactive oppositions generating these questions are not formally differentiated from the dichotomies that govern subsequent episodes in the narrative. In these episodes, as in the Happy Valley, the contrast between pastoral species and beasts of prey is paralleled by the inner opposition between pleasures and pains, 'blessings' and 'evils', 'repose' and 'discontent'. On another level, the same paradigm is easily converted into a variation of the by-now familiar Johnsonian antithesis of masters and slaves: a contrast between warriors, brigands, and princes on the one hand, and hermits, shepherds, sages, and scientists on the other. It can be further extended to the contrast between male activity, defined by hunting and fighting, and female 'diversions', which, in the sheltered world of the Arabian camps, are restricted to what Nekayah's confidant, Pekuah, calls 'childish play'. The whole complex, finally, is embodied on the intellectual plane by the contrast between an imagination which 'preys' incessantly on life and a reason which pacifies, socializes, and generally counsels restraint.

The attributes that serve to make up this series of oppositions are not always kept symmetrical in the world outside the Happy Valley, but sometimes appear in complex combinations, permutations, and reversals. Thus Arabia is described by Imlac, for example, as 'a nation at once pastoral and warlike; who live without any settled habitation; whose only wealth is their flocks and herds; and who have yet carried on through all ages an hereditary war with all mankind, though they neither covet nor envy their possessions' (9, 60). Here the union of pastoralism and martial strife does not lead to the image of a society that is hermetically sealed, like Abyssinia but one which, though beset by a similar contradiction, is driven outward rather than inward. The Arabs are portrayed as nomadic tribesmen who live exclusively by raising sheep and oxen and yet practice a kind of motiveless war that implies a total reversal of the values upon which their pastoral culture rests.

Since the same oppositions can occur within the mind of the individual, they can also engender the same contradictions. The Arab chieftain who seizes Pekuah is an obvious case in point, since he exemplifies within his character the same ironies that characterize his people. A fierce brigand whose 'revenue is plunder', the Arab chieftain wages unceasing warfare on neighbouring kingdoms. But while he subscribes to the belief that 'the violence of warfare admits of no distinction', he is also a civilized man who observes 'the laws of hospitality with great exactness to those who put themselves in his power'. This disclaimer can be seen as consistent with the chief's reiterated stress on his own humanity. But is also serves the more devious purpose of allowing him to extend the identification of civility with subordination from sheep and oxen to women and slaves. The dualism that is obliterated in warfare is thus regenerated in peacetime, embodied in the rigid segregation of men and women, masters and slaves. It is significant in this connection that the Abyssinians who are pastoralists without being predatory do not practice this kind of segregation by gender.

The superiority of men to women is of course deeply ingrained in the Arabic culture of the narrative. It is in line with the principle that elevates 'wild' nature over the tepid conventions of art. Yet there is a very real sense in which the Arab chieftain's growing attachment to Pekuah testifies to the falsity of this opposition. The Arab women, who exhibit the same discontent as the inhabitants of the Happy Valley, are compared to caged birds and timid lambs – weak, dependent creatures that resemble the sheep and oxen the Arab men tend. Prevented from rising above the plane of animal existence, they have obviously failed to develop the 'powers of reflection, judgment, and ratiocination' that alone can make poetry and enlightened conversation possible (*Letters*, II, 79). Instead they are portrayed as being imprisoned within a world of nominalistic particulars, passing part of their time 'in watching the progress of light bodies that floated on the river, and part in marking the various forms into which clouds broke in the sky' (39, 130). Yet this vacancy is precisely what renders them unfit for the company of a man who, in keeping with the real pastoral values of his culture, treasures domestic felicity, yet insists upon wielding the kind of predatory power which renders that felicity impossible. By deriving 'the rules of civil life' from the very same will-to-domination he exercises in wartime, the chief negates the very purpose these rules were meant to serve.

*   *   *

Different permutations of the same system of oppositions organize the description of the successive choices of life that Rasselas and Nekayah consider. These permutations reintroduce the same antitheses of inside and outside, pastoralism and warfare, blessings and evils, that governed the classification of animals within the Happy Valley. In the political realm, the application of these antitheses necessarily engenders the same fortress mentality that brought the Happy Valley into existence. The wall of China is a signal instance of the kind of boundary this mentality erects in its effort to preserve itself from the dangers of external invasion:

> Of the wall it is very easy to assign the motive. It secured a wealthy and timorous nation from the incursions of barbarians, whose unskillfulness in arts made it easier for them to supply their wants by rapine rather than by industry, and who from time to time poured in upon the habitations of peaceful commerce, as vultures descend upon domestic fowl. Their celerity and fierceness made the wall necessary, and their ignorance made it efficacious.   (33, 113)

The problem with such efforts to keep out 'the incursion of barbarians' is that they invariably prove incapable at some point of dealing with the barbarians within. The uniting of men behind walls necessarily involves the exercise of political power and authority. Inevitably subordination is introduced and with it a cruelty that can lead to a reversal of the opposition between inside and outside and a collapse of the boundary dividing them:

no form of government has been yet discovered, by which cruelty can be wholly prevented. Subordination supposes power on one part and subjection on the other; and if power be in the hands of men, it will sometimes be abused. The vigilance of the supreme magistrate may do much, but much will still remain undone. He can never know all the crimes that are committed, and can seldom punish all that he knows. *Rasselas wants himself a perfect governor* (8, 55)

The roles of magistrate, subject, and criminal almost seem here to lose their distinctness, their representability, which is what makes this passage so fitting a commentary on the various images of enclosure in *Rasselas*.[4]

Similar complications act to undermine the idealized projection of Johnson's paradigmatic oppositions in their explicitly asserted form as choices of life. The gardens of the exiled prince are constructed upon the same 'artificial' principle of exclusion as the Happy Valley:

> The shrubs were diligently cut away, to open walks where the shades were darkest: the boughs of opposite trees were artificially interwoven: seats of flowery turf were raised in vacant spaces, and a rivulet, that wantoned along the side of a winding path, had its banks sometimes opened into small basins, and its stream sometimes obstructed by little mounds of stone, heaped together to increase its murmurs. (20, 86)

Yet the prince's effort to construct such an artificial paradise, like the similar effort of the Abyssinian rulers, is doomed to failure, for the prosperity which makes such an effort possible inevitably reintroduces, as one of its preconditions, the very insecurity it sought to conquer. As the prince succinctly puts it, 'my condition has indeed the appearance of happiness, but appearances are delusive. My prosperity puts my life in danger' (20, 86–7).

The hermit offers an example in which the same antithesis of the pacific and predatory recurs but where the source of the failure is spiritual rather than material. In terms of this antithesis, the hermit's life is divided into an early period of martial activity and a late period of pious inaction. In the early period, 'the hermit professed arms and was raised by degrees to the highest military rank' (21, 88), while in the late period he flees to the shelter of a quiet cavern, employing 'artificers to form it into chambers' and storing it 'with all that I was likely to want' (21, 89). This, of course, is precisely the same process of enclosure and division to which the Abyssinian 'artificers' conformed in erecting the Happy Valley. Yet if the hermit has thus 'escaped from the pursuit of the enemy' without, he has rediscovered the enemy within. This enemy is embodied in the 'thousand perplexities of doubt and vanities of imagination' that prey upon his mind. Prevented by these perplexities and vanities from exercising his rational capacity for generalization and classification, he is condemned, much like the Arabian ladies, to the empty nominalistic enterprise of 'examining the plants which grow in the valley, and the minerals which' he collects 'from the rocks' (21, 89).

The example of the hermit indicates that other types of opposition may intervene in the succession of individuals and groups who confront Rasselas and Nekayah with the various choices of life they must consider. Oppositions may be established on either the material level or the intellectual level: the same figure of the predator functions at different levels of abstraction in the narrative. Again, there are oppositions between wild and tame (the contrast between the young men of spirit and gaiety on the one hand and the daughters on the other); or between the high or 'great' (the stoic sage and exiled prince) and the low or 'mean' (the young men and shepherds). In the latter paradigm, greatness refers either to wealth or knowledge, meanness to poverty or ignorance.

An analysis of these oppositions is not intended to produce a strikingly new interpretation of the successive choices of life that offer themselves to Rasselas and his sister. What it does reveal is the underlying consistency of the 'artificial' perspective governing many of these choices. Nor is this necessarily a matter of academic interest, for it is never merely an external agency that produces the failure of each choice, but the internal contradictions and reversals within the play of oppositions that generated it. Thus the highly artificial philosophy of the 'wise and happy man' is undermined not only by the natural grief he expresses at the loss of his daughter but also by the contradiction inherent within the philosophy itself. For it is the same 'reason and truth' which teaches him that 'happiness is in everyone's power' that also tells him that his 'daughter will not be restored' (18, 84).

Rasselas's response to the wise and happy man conveys the extent of Johnson's scepticism about the philosopher's 'reason and truth'. To discover the contradiction within such a philosophy is to disclose its detachment from reality: an argument that purportedly serves as a guide to life manifests itself, Rasselas learns, as purely phonic, as a discourse which is composed of 'rhetorical sounds', of 'polished periods' and 'studied sentences'. In this way, philosophic wisdom becomes emptied of meaning: its purely illusory logic is shown to be nothing more than the seductive art of rhetoric. What is in question, for Johnson, is not an alternative wisdom but the verbal ingenuity of a wisdom which preserves itself by concealing the emptiness of its tropes and figures behind a façade of voice as presence. To discover this emptiness is to learn what Rasselas learns when he listens to the sage who advocates a life according to nature and discovers that he 'was one of those sages whom he should understand less as he heard him longer'. (22, 92).

The potential aberrations of rhetoric are evoked in *Rasselas* not only in the studied cadences of the philosophic sage but, to a certain extent, in the utterances of other characters including the astronomer, Rasselas, Nekayah, and even Imlac. Behind the obvious similarities in tone, imagery, and diction lies a shared indebtedness to the tropes and figures of persuasive rhetoric. Of course this indebtedness does not produce the same degree of error and illusion in all of the different arguments advanced in the narrative, any more than it insures their adherence to reason and truth. Rather, it marks a divergence that is seen to inhere within all persuasive discourse, an ever-present temptation to phonocentric illusion and temporal wish-fulfilment that renders its relation to reality problematic. Johnson's method of eliciting truth in *Rasselas* from the carefully contrived encounter of question and answer, declaration and interruption is one way of indicating this divergence. Again and again, its effect is to draw attention to the rhetorical nature of whatever is being affirmed.

One might say, thus, that Rasselas participates both in the rhetorical conventions of philosophical dialogue and the moral-psychological mode of narrative. It goes beyond the kind of illustrative performance exemplified in the Platonic dialogues: but it does not invite the reader to consider its dramatic interchanges merely as fictions.[5] Thus life is not only an external phenomenon providing an empirical test for the psychological truth of the arguments. It is also visible within the narrative as a topic which, by its recurrence, traces and delimits the horizon of these arguments. While avoiding the fallacious illusion of logical demonstration, the interchanges of the characters yet retain the possibility of arriving at productive insights and even the closure and finality of aphorism: conversely as products of rhetorical discourse, these insights are always considered provisionally; incessantly parenthesized by question and answer, the moral affirmations of *Rasselas* also emerge as indicators of the lexical process generating them.

It is the duplicity to which such a process is potentially subject that links the orations of the philosophic sages to the pleasure palaces and gardens of the architects and engineers in *Rasselas*. Just as the studied cadences of the sage are meant to lull the 'pure' mind into a state in which it will become immune to grief, suffering, or earthly desire, so the habitations of the architect are designed to insulate the community from the 'calamities' of the outside world. Values, beliefs, and inventions are seen as techniques that will promote either virtue – the domain of the sage – or pleasure – the province of the 'artist'. But where the artfully modulated sentences of the sage are susceptible to doubt, the mechanical inventions of the artificer are liable to failure. There are three such instances of failure in *Rasselas*, each one promising the same command over nature as the rules of the sages: the flying 'wings' of the mechanic artist; the controlled weather of the astronomer, and the pyramids of the pharaohs. The 'artist' who constructs the ill-fated wings for Rasselas has already won considerable eminence for his mechanical marvels:

> By a wheel which the stream turned, he forced the water into a tower, whence it was distributed to all the apartments of the palace. He erected a pavilion in the garden, around which he kept the air always cool by artificial showers. One of the groves, appropriated to the ladies, was ventilated by fans, to which the rivulets that ran through it gave a constant motion: and instruments of soft music were placed at proper distances, of which some played by the impulse of the wind, and some by the power of the stream.   (6, 49–50)

Equally seductive as the periods of the sages in its auditory appeal, this hydraulic technology is constructed on the principle of the closed cycle: the marvels thus described are the local effect of a general economy characterized by the incessant circulation of a flowing that has no goal or termination. But such a 'constant motion' is subject to destructive alteration. The unbearably repetitive 'soft music' excluding temporal succession or spatial variation, will inevitably grow 'tasteless and irksome'.

There is a sense in which this mode of technology is susceptible to the same critique as the rhetorical art of the sage. Both assume that it is possible to ignore human limits in seeking to escape from pain and discomfort. Thus the artist's wish 'to tower into the air beyond the malice and pursuit of man' leads him to ignore the fact that 'every animal has his element assigned to him: the birds have the air, and man and beasts the earth' (6, 50), and the disaster which befalls his machine is an indication of the extent to which it has become divorced from natural laws and processes. The astronomer's delusion that he can control the weather, though initially less credible, nonetheless embodies in an intellectual realm what the artist has attempted to achieve in the material realm.

Moreover, both astronomer and artist become isolated from their fellow creatures by their wish to flee from the 'pursuit of man'. Indeed, it is their self-imposed isolation that contributes to their downfall. Working outside the boundaries of the human community, both have started from limited techniques and material resources and proceeded insensibly to a 'super-technology' of total domination that ensures their elevation in a skyworld of error and illusion. Yet however spectacular are the engineering fiascos of the artist and astronomer, they possess the plausibility of projects that are at least aimed at promoting 'the universal good'. When the mind becomes disengaged from such a goal, it becomes subject to the imagination that preys incessantly on life. Such grandiose 'follies' as the pyramids of Egypt embody the same soaring will-to-power as the artist's wings, but now this power is harnessed to a scheme that is 'to no purpose' and 'without end'. Divorced from any concrete aims, the pyramids have become the most dazzling instance of the law that 'he that built for use, till use is supplied, must begin to build for vanity, and extend his plan to the utmost power of human performance, that he may not soon be reduced to another wish' (32, 113–14).

*   *   *

The conception of art underlying the Happy Valley and the subsequent choices of life thus encompasses a polarity between inside and outside, a notion of the purity of taxonomic boundaries, and a belief in the ready availability of a technology that will protect the self from external threats. As such, this technology is largely defensive, drawings its inspiration from the belief that man possesses a real power over the realms of nature and mind and justifying this belief by the assumption that man is linked by the principle of analogy to the animals below (the sphere of pleasure) and the angels above (the sphere of virtue) in the scale of nature. Indeed, it is man's very access to power that, in this way of thinking, enables him to organize existence in such a way as to satisfy the desires and aspirations that link him to these two orders of being.

This of course is the point at which this conception of art is subverted and displaced by a different and alternative perspective. The basic assumption of this perspective is first presented in Rasselas's meditation on life in Chapter 2. Because existence in the Happy Valley is devoted more to the pursuit of pleasure than virtue. Rasselas's meditation takes as its starting point his observation on a specific group of animals:

Rasselas, who knew not that any one was near him, having for some time fixed his eyes upon the goats that were browsing among the rocks, began to compare their condition with his own.

'What', said he, 'makes the difference between man and all the rest of the animal creation? Every beast that strays beside me has the same corporal necessities with myself: he is hungry and crops the grass, he is thirsty and drinks the stream; his thirst and hunger are appeased, he is satisfied and sleeps; he arises again and is hungry; he is again fed and is at rest. I am hungry and thirsty like him, but when thirst and hunger cease I am not at rest; I am like him pained with want, but am not like him satisfied with fulness. . . . Man surely has some latent sense for which this place affords no gratification; or he has some desires distinct from sense, which must be satisfied before he can be happy'.(2, 41–2)

Although Rasselas echoes traditional denials of the continuity between men and animals, he also articulates what seems to be a distinctive sense of desire as at once a lack, or a want, that cannot be satisfied and an incessant, if latent, movement toward fulfilment. This movement can never be quelled and, for that reason, can never be enclosed within a prison of binary, reactive oppositions or within a taxonomic scheme that asserts the natural unity of the angelic, human, and animal realms. The conclusion of Rasselas's meditation might therefore be the affirmation of an inside that is already a difference, already an outside.

The sense of discontinuity evoked by Rasselas's discovery of desire and generic anomaly recalls Johnson's critique of Jenyns's *Inquiry*. In the traditional conception of the scale of being, every species is at once linked to and differentiated from all other species by its capacity to achieve its own particular happiness. But insofar as human desire is unappeasable, it can never serve as the norm for a distinctively human happiness. However much the human mind seeks the kind of felicity enjoyed by all other beings, this felicity remains unattainable. As Imlac later puts it, desire is either fulfilled by its transformation into its opposite, satiety, or it remains an ungratifiable 'appetite for novelty' (32, 114). There can thus be no real opposition between desire and aversion, pleasure and pain, good and evil, since there is no certain way of distinguishing between them. In insisting that 'the causes of good and evil . . . are so various and uncertain, so often entangled with each other, so diversified by various relations, and so much subject to accidents which cannot be foreseen' (16, 79), Imlac testifies to this uncertainty. The term 'cause' becomes transformed in his analysis from a necessary link in the natural order to an event that, for all practical purposes, is not accessible to human reason.

The ungraspability of causes does not render choice impossible but does put in question the validity of the view that the consequences of a particular choice are easily ascertainable.

Thus the entire problem of causation, like that of continuity, is displaced in what might be called a 'sceptical' fashion. Order has disappeared, as much in the natural world, as in the human world. In place of a vision of organic unity and plenitude, Imlac posits a different kind of totality, the totality of an 'inexhaustible variety', a non-unified, heterogeneous plurality. The act of contemplating this plurality is for Imlac a necessity since it aids one in alleviating 'the gloom of perpetual vacancy' within and the prospect of 'barren uniformity' without. Against the horizon of this vacuity, differences between phenomena take on a special importance: 'the only variety of water is rest and motion, but the earth has mountains and vallies, deserts and cities; it is inhabited by men of different customs and contrary opinions' (9, 58). Imlac analyses the activity needed for a poetry that takes these differences into account as a twofold process. First, through the act of marking individual images, the poet apprehends variety as novelty, as the careful observation of what has not-yet-been seen. To avoid being overwhelmed by 'the different shades in the verdure of the forest', however, the poet must also introduce identity into difference, general representation into nominalistic particularity. But resemblance does not refer in this theory to a system of correspondences linking species to an organic whole. Similitude without correspondence refers, rather, to 'the prominent and striking features' within species. Insofar as the task of the poet is to 'remark' these features, it is to re-present them in order to re-discover their 'original' as a repetition of the same. It follows that for the poet the unknown now becomes the not-yet-recognized known, that to learn is now to remember. Yet while this theory of representation locates the challenge of poetry in the suppression of 'minuter discriminations' within classes, it does not erase all difference, for there is an equal and implicit challenge in the need to preserve differences between classes.[6]

This poetry of difference without nominalistic particularization and large appearance without correspondence thus works to exclude the possibility that man can ever be fully comprehended link in a hierarchically ordered chain of beings. Yet the discontinuities that prevent man from discovering analogies between the different parts of the natural creation do not keep him from learning from other 'animal species'. Denied a particular niche of his own, man is free to borrow and copy from species which inhabit such niches. The simple technology which arises from this mode of imitation differs strikingly from the technology of the 'mechanic artist'; though it requires an implement in order to 'make' something, it adheres closely to natural processes and natural laws. In this respect, Johnson shares the prejudices of the eighteenth-century artisan class into which he was born. Not only does he find the mechanical arts to be equal in value to the liberal arts, but the activities of the skilled or semi-skilled craftsman are worth more, in his eyes, than the grandiose projects of the engineer or architect. Accordingly, he takes a particular interest in the level of technical expertise that he later finds in the agriculture and handicrafts of the Highlands and Western Islands of Scotland. But in *Rasselas*, Johnson's enthusiasm for this artisan technology is not experienced as historical; indeed, it is supported by an appeal to nature and assumed to be universal rather than particular and artificial. Thus Imlac and Rasselas, instead of trying to construct 'wings' that defy the laws of gravity, find a mode of transportation that depends upon burrowing rather than flying:

> As they were walking on the side of the mountain, they observed that the coneys, which the rain had driven from their burrows, had taken shelter among the bushes, and formed holes behind them, tending upwards in an oblique line. 'It has been the opinion of antiquity', said Imlac, 'that human reason borrowed many arts from the instinct of animals: let us, therefore, not think ourselves degraded by learning from the coney'. (13, 72)

This activity has more than a local and strictly narrative purpose: it derives from the same painstaking process of observation that Imlac insists is central to his vocation as a poet. The relations which this process is designed to discover or construct are most commonly based on contiguity (the function of burrowing common to both the coney and the miner). Such relations do not imply a genetic resemblance but are rather based on the possession by beings of one or more common characteristics. Thus they differ from the comments of the sage who urges men to 'observe the hind of the forst, and the linnet of the grove: let them consider the life of animals whose motions are regulated by instinct: they obey their guide and are happy' (22, 91). This is the perspective of excessive generality and abstraction. Proposing an all-encompassing formula of life in harmony with nature, it considers only the genus 'animal' and overlooks the difference between 'species'.

*   *   *

There is one other alternative to the inward-looking technology of the engineers who are responsible for the construction of the Happy Valley, the pyramids of Egypt, and the great wall of China. This is a technology in which the outside becomes inside, predatory warfare is transformed into heroic enterprise, and conflict reemerges as communication. Imlac finds these characteristics implicit in the enterprise of 'the northern and western nations of Europe':

> they cure wounds and diseases, with which we languish and perish. We suffer inclemencies of weather which they can obviate. They have engines for the despatch of many laborious works, which we must perform by manual industry. There is such communication between distant places, that one friend can hardly be said to be absent from another. Their policy removes all public inconveniences: they have roads cut through their mountains, and bridges laid upon their rivers. And, if we descend to the privacies of life, their habitations are more commodious, and their possessions more secure. (11, 66 – 7).

What is important here is not merely the dramatic difference in the level of material culture, but the accompanying difference in the thrust of the technology, from building walls and enclosures to cutting through mountains and erecting bridges. This technology almost seems to anticipate the great projects of the next century, projects that involve the breaching of the earth for the opening up of communication and transportation. That Rasselas sees the possibility inherent in this technology accounts for his exclamation that 'they are surely happy . . . who have these conveniences'. But Imlac, who knows better, holds simply that 'the Europeans are less unhappy than we, but they are not happy' (11, 67).

Imlac's refusal to acknowledge the possibility of a system based on a reversal of the priorities that govern existence in the Happy Valley helps explain why the conflict between these terms gives rise on the narrative plane not to a resolution but an impasse. In the case of the hermit, for example, the clash between the timorous and the predatory generates a potentially endless series of reversals, between 'society' and 'solitude': the spectacle of cruelty and injustice leading to a flight into solitude; the awareness of an internal struggle between the equivalent passions prompting a return to society. Yet this mode of reversal, to which the hermit succumbs, is not arrested by the return to society, for society, like its opposite, reveals an essential lack within it, so that solitude becomes a necessary supplement to what has just replaced it. At least one bystander, recognizing the implications of this reconstitution of desire, argues that it was likely:

> that the hermit would, in a few years, go back to his retreat, and, perhaps, if shame did not restrain or death intercept him, return once more from his retreat into the world. (22, 90).

The apparently unending nature of this play of reversals is seen most clearly in the famous decision of the travellers to 'return to Abysssinia', a decision which, like the change of mind of the hermit, could itself be subject to potential reversal. It is also apparent in the debate between Rasselas and Nekayah over marriage. The details of this debate are noteworthy, for they show how the solitude/society polarity gives rise to a particular kind of closure. Unlike modern anthropologists, Johnson shows no interest in viewing marriage as a link between the family and society. Nor does he show any inclination to stress the importance of biological ties in marriage. Indeed it is clear that like some modern commentators Johnson views the process by which we fall in love and plight our troths as a randomizing one, in which we are often matched with a person who is inappropriate:

> a youth and maiden, meeting by chance or brought together by artifice, exchange glances, reciprocate civilities, go home, and dream of one another. Having little to divert attention, or diversify thought, they find themselves uneasy when they are apart, and therefore conclude that they shall be happy together. They marry, and discover what nothing but voluntary blindness before had concealed; they wear out life in altercations, and charge nature with cruelty. (29, 105).

The consequences of this view of the weak bonds between husband and wife is to direct our attention to the extended family. To describe the internal structure of the family, Johnson draws upon a traditional analogy between kin groups and political kingdoms: 'if a kingdom be, as Imlac tells us, a great family, a family likewise is a little kingdom, torn with factions and exposed to revolutions' (26, 97). This conception of the family can be traced back to patriarchalist theories of the origin of political institutions that dominated political thought in the seventeenth century.[7] In Johnson's late, attenuated version of the analogy, the force that insured the authority of the patriarch has been overthrown, and rivalry and revolution are allowed to reign unchecked. The principal parties in this internecine warfare are parents and children, parties whose interests are coterminous with the same principle that divides all society into warring factions:

> parents and children seldom act in concert: each child endeavours to appropriate the esteem or fondness of the parents, and the parents, with yet less temptation, betray each other to their children; thus, some place their confidence in the father, and some in the mother, and by degrees the house is filled with artifices and feuds.   (26, 97)

Here, the overthrow of patriarchal authority has led to a reversal in which the parents are timorous and the children the aggressors. The inherent instability of the family is not overcome by the alleged superiority of marriage (the equivalent of society) to a life of celibacy (the equivalent of solitude), and a further division of early and late marriages is introduced as a means of overcoming the contradictions that have been disclosed. But the same pattern of reversals that governs the choice between society and solitude extends itself to this choice, and the two alternatives resolve themselves into yet another version of the same destructive dualism: the early marriage displays the same internecine warfare that characterizes the family in the earlier accounts; the late marriage fails to overcome the inertia of habit and custom that characterizes celibacy. The metaphor of the 'source' and 'mouth' of the Nile, which Imlac introduces, only partly resolves the dilemma, since it implies that the issue is finally undecidable. Thus it is reduced to asserting that we must make a choice, even though we lack the wisdom of deciding which of the two alternatives should be given priority over the other.

It is on the basis of this imperative that the play of reversals is at least temporarily stopped. But this peculiar fiat, this arbitrary and contingent choice is neither assertoric nor categorical in the Kantian sense. It affirms neither that we must act in our best interests nor in such a way that our action could become a law for others. Asserting that motion is preferable to rest, activity to inaction, it implies that we must be prepared to accept the concatenation of good and evil, pleasure and pain that will accompany any choice. In terms of narrative convention, the equivalent of this logic of undecidability is 'a conclusion in which nothing is concluded'. Indeed, the abruptness of Johnson's conclusion is in no way a conclusion: it is nothing more than a stroke of the pen, arbitrarily choosing its own termination. If the narrative's conclusion is organized around its failure to conclude, that conclusion turns out not to be a teleological limit inherent in the text – its true end – but an arbitrary stopping point, an enactment of the principle of undecidability that governs the imperative to choose between celibacy and marriage, or between early and late marriages.[8]

It might be argued that the rediscovery of continuity implied in the metaphor 'the stream of life' might provide an alternative to the anti-categorical imperative implied in this logic. The paragraph that establishes the plans of Imlac and the astronomer seems to suggest a new source of plenitude:

> Imlac and the astronomer were contented to be driven along the stream of life without directing their course to any particular port.   (44, 158)

Yet this metaphor is not immune to criticism. It has its own interior logic: a temporalization that escapes termination by delighting in the flow of its own motion. This flow, grounded (like the 'rest and motion' of the ocean) in a repetition of the same, can neither generate change, nor account for it. But a complicating tendency is implied in the phrase 'particular port' which is also a surreptitious introduction of the dialectic of desire and satisfaction into a metaphor which had seemingly banished it. Far from having achieved a new link to plenitude and the present moment, Imlac and the astronomer may have only discovered another variation of the same process of deferral and delay evident in a failure to choose between early and late marriage. The inevitable consequence of this failure lies in the recognition that one's 'acquirements are now useless' (12, 69–70). The abandoned aspirations of Imlac and the astronomer are perhaps the most prominent evidence that the stream of life is actually a choice of life, subject to the same illusions and limitations that govern the illusory choices of Pekuah, Nekayah, and Rasselas.

If this interpretation has any validity, it suggests that two entirely coherent but incompatible readings can be imposed on the same passage. To be sure, the occurrence of such an ambiguity within the confines of a brief paragraph obviously does not have the same effect in a tentative concluding chapter as in a more extended narrative. On the other hand, one may well wonder, with equally good reason, whether the pattern of ambiguity does not represent a more faithful interpretation of Johnson's intention than some of the totalizing alternatives, simply because the metaphor may conceal incompatible affirmations.[9] It is also consistent with the text's deliberately unresolved contradictions of viewpoint, its willingness to suspend or reverse the earlier and supposedly authoritative choices of Imlac and the hermit. The title of the concluding chapter can in this case be taken to imply the holding together of two possibilities without the least need of choosing between them. Yet this perspective would of course be akin to that of Imlac and the astronomer insofar as it repudiates the imperative implicit in the need to choose between an early and late marriage. Yet the 'conclusion in which nothing is concluded' is intended precisely to transcend or discredit any such facile interpretations. What it suggests is that Johnson controls the rhetoric of plenitude and presence in all 'choices of life', from the most naïve to the most devious forms, including even the refusal to make a choice of life.

The affirmation of a 'choice of eternity' founded in the transcendental valorization of the dichotomies of pleasure and pain, security and danger, represents, of course, a considerable departure from the epistemological and rhetorical critiques to which the earlier choices of life were subjected. To ignore this choice, or to discount its value, as some have done, is to ignore the seriousness with which its arguments are advanced.[10] But the choice of eternity can also be called into question, since its reliance upon the same system of oppositions as the earlier choices can raise the same objections. Johnson counters this likely objection by making it clear that the choice of eternity is one that everyone must confront. Thus, he implicitly rejects the notion that the choice of eternity belongs to the same chain of successive errors as the choices of life to which it is opposed. On the other hand, the state of mind appropriate to the choice of eternity can only come about on condition that the individual has already achieved an adequate degree of 'mortification'. The individual who has committed himself to life, the world, and to society may find it difficult to achieve such a state of enlightenment. Hence it is not surprising that Nekayah's impassioned declaration is linked in the last chapter to the same kind of illusory 'scheme of happiness' that the narrative has been at such pains to discredit:

> the princess thought, that of all sublunary things, knowledge was the best: she desired first to learn all sciences, and then proposed to found a college of learned women, in which she would preside, that, by conversing with the old, and educating the young, she might divide her time between the acquisition and communication of wisdom, and raise up for the next age models of prudence and patterns of piety.   (49, 158)

The choice of eternity is thus represented as a kind of climax, but one which begins and ends, it appears, within a kind of 'religious' parenthesis. Only in this way can *Rasselas* sustain its narrative coherence and ironic strategies.

Johnson's narrative therefore refuses to make the choice of eternity into the central event in the lives of its characters. For all its decisive significance, the episode is relegated to a secondary status where it is not allowed to resolve the inner contradictions of the narrative. It is these contradictions which offer perhaps the best insight into the divergent perspectives that inform the design of *Rasselas*. In the opposition between art and nature, inside and outside, solitude and society, the second term has for its aim to reveal the existence of hidden impurities and complexities within the supposedly pure and distinct worlds designated by the first term. Yet *Rasselas* thwarts an easy interpretation of this purpose by refusing, in the last analysis, to commit itself to either set of terms. Instead, it dwells upon the discontinuities, the unmasterable limitations of both sets, and the way they evade the norms upon which any choice of life must be based.

7

The Anthropology of Natural Scarcity in *A Journey to the Western Islands of Scotland*

Part of the continuing fascination of Johnson's *A Journey to the Western Islands of Scotland* may lie in its singularity.[1] It asks to be read not as a conventional topographical guide but as an extended essay in cultural investigation and interpretation. As such, it is probably a companion to the theory of travel writing that Johnson put forth more than a decade earlier in *Idler*, No. 97. In that essay, Johnson attacks what he regards as the predictable narratives and generalized decor of most travel books and advocates the careful description of men and manners.[2] Travel writing emerges in his view as the product of a differential analysis in which knowledge is acquired through the comparison of different cultures. Elaborating on this principle, Johnson holds that:

> every nation has something peculiar in its manufactures, its works of genius, its medicines, its agriculture, its customs, and its policy. He only is a useful traveller who brings home something by which his country may be benefited; who procures some supply of want or some mitigation of evil, which may enable his readers to compare their condition with that of others, to improve it whenever it is worse, and whenever it is better to enjoy it. *(Idler, Adventurer*, 300)

Seen in the light of this programme, a travel-narrative becomes a comparative analysis of a culture as a distinctive entity. Yet the extent to which *A Journey to the Western Islands* actually carries out such a programme has been much debated. The opinion, common among Johnson's contemporaries, that it was an attack on Scotland and the Scotch has been recently revived by Patrick O'Flaherty.[3] More subtle and sympathetic critiques are advanced by many critics, among them, Patrick Cruttwell and Donald Greene. In a fascinating essay on changing attitudes toward the Highlanders in the eighteenth century, Cruttwell points out that the reviewer in *The Gentleman's Magazine* took Johnson 'to task for being altogether too well disposed to the rebels of the '45' and concludes that on the issue of the Conquest Johnson's mind was 'utterly divided, totally uncertain'. Greene shows how strongly Johnson's whole approach anticipates the techniques and presuppositions of modern cultural anthropology. Johnson, writes Greene, was 'tremendously interested' in how the Highlanders 'actually *live*' and has a capacity for 'empathy' which enabled him to put 'himself in the position of the individual Highlander'.[4]

The source of this dispute undoubtedly lies with Johnson himself. His expectations, antipathies, and contradictory motives make it difficult to assess his *Journey* with any degree of confidence. Nonetheless, there are at least two senses in which *A Journey to the Western Islands* can be legitimately viewed as a work of cultural anthropology. One sense is general: its field of study is an alien and relatively backward people. The other sense is highly specific: its basic assumption is that the totality of this people's customs form an ordered whole, a system.[5] Indeed, Johnson actually uses the term 'system' several times, referring variously in the *Journey* to 'the whole system of things' (*Journey*, 10), an old woman's 'whole system of economy' (*Journey*, 33), and the 'system of antiquated life' (*Journey*, 57). In each of these usages, Johnson is referring to a perspective in which the constituents of cultural behaviour – ceremonies, kinship relations, political institutions, methods of farming and cooking – are seen, not merely as isolated and discrete empirical entities, but as parts of a larger pattern. That Johnson has something like this very much in mind is suggested by the attention he devotes in certain sections of the text to a systematic enumeration of the 'peculiar' manners of the Highlands and Western Islands.[6] In *The Dictionary*, Johnson defines the word 'peculiar' as 'the exclusive property' of something and as 'belonging to one, to the exclusion of others', and his usage here suggests that the 'something peculiar' is what unites elements as diverse as manufactures, medicines, agriculture, and customs into a unified picture.

Viewed in the light of this assumption, Johnson's outlook can be seen as resembling that of the nineteenth-century British anthropologist, E. B. Tylor, who, in a famous passage, defined culture as 'that complex whole which includes knowledge, belief, art, morals, law, custom, and any other capabilities acquired by man as a member of society'.[7] As the exemplification of this perspective, *A Journey to the Western Islands* is probably more akin to a number of eighteenth-century studies of the origins of human culture than to the travel writings of Johnson's predecessor in his trip to Scotland, Thomas Pennant. These studies include Montesquieu's *L'Esprit des Lois* (1748), Rousseau's *Discours sur l'Origins d'inegalite* (1759), Adam Ferguson's *Essay on the History of Civil Society* (1766), and John Millar's *Origin and Distinction of Ranks* (1771). Like these early anthropological texts, *A Journey to the Western Islands* pictures society as a system of interdependent parts rather than an aggregation of novelties and topographical sights and sounds; and, like them, it relates each society to a generalized pattern of social development, a series of stages through which all societies must pass.[8] The relation of Johnson's travel memoir to these ancestors of modern anthropology has been somewhat obscured by its decidedly empirical, even sceptical attitude to what might constitute acceptable evidence. Unlike a Montesquieu or a Ferguson, Johnson bases his generalizations about human culture on an actual journey to an alien, relatively backward world.

The result of Johnson's scepticism is a book that comes ultimately to resist the systematic closure evident in eighteenth-century ventures in speculative anthropology. Thus Johnson never assumes that it is an easy task to 'decode' the culture of the Highlands or Western Islands in order to reveal an underlying unity. The fragmentary character of *A Journey*, Johnson's mistrust of oral tradition, awareness of historical change, and scepticism concerning the veracity of even the best informants – these attitudes may be seen as strategies designed to frustrate theoretical foreclosure and too easy appropriation of the 'other'.[9] We may be perceiving in the form of his work an implicit critique of an anthropology that assumes that every country embodies a functionally integrated way of life and emphasizes the harmonizing, stabilizing, and socially supportive effects of cultural patterns.[10] Johnson, by contrast, portrays a society that seems inherently degenerate and gives full emphasis to the disruptive, disintegrative, and psychologically disturbing aspects of life in the Highlands and Western Islands.

A Journey to the Western Islands thus resembles a work of cultural anthropology in its emphasis on the 'peculiar' character of individual nations, but its form does not so much resemble a systematic treatise as a series of discrete encounters, notes, and set pieces that enact the movement of Johnson and Boswell through 'regions of barrenness and scarcity'. Its design is in the form of a heroic quest – an implied departure of Johnson and Boswell from a homeland grown familiar and comfortable; their journey into a world that grows progressively more alien as they leave the more populous south for the Highlands and Western Islands;

their encounter with surprises, real and imaginary dangers, challenges and discoveries; the assistance they receive from a variety of strangers; and their implicit return with a deepened knowledge of reality and the need to communicate what they have learned to those who, less adventurous, have stayed behind.[11] Asking what sort of person dwells in 'the penury of these malignant regions', Johnson seeks to reconstitute the social organization, political institutions, and beliefs that from beneath the surfaces of Scottish life, might possibly have produced and sustained that penury.

Yet to comprehend these elements is also to recognize once again the limitations of the principle of plenitude. Indeed, it is possible that Johnson's fascination with a society dependent on conditions of natural scarcity may have been shaped by expectations that grow out of his critique of Jenyns's *Inquiry*. The view that the natural world is characterized by a boundless abundance and variety may be one of the casualties of this critique. Such an abundance and variety no longer have any philosophical justification, as George Cheyne may have recognized when, in his *Essay on Regimen*, he openly declared that:

> Our earth has, in particular, *barren and unfruitful Climates, unhospitable and uninhabitable Regions, unhealthy and mortal Seasons, Tempests, Thunder and Lightning, Volcanos, Plagues and epidemical and infectious Atmospheres.*[12]

For his part, Johnson was clearly convinced that the barrenness of Scotland was not a regional aberration but part of a much broader topographical phenomenon:

> Regions mountainous and wild, thinly inhabited, and little cultivated, make a great part of the earth, and he that has never seen them, must live unacquainted with much of the face of nature.   (*Journey*, 40)

It is perhaps in the context of this perspective that Johnson's preoccupation with the lives of a people dwellling in a habitat that does not naturally and spontaneously regenerate itself needs to be seen. In a sense, this preoccupation is the flip side of Johnson's concern with temporal experience in the *Rambler* and *Idler* essays. In seeking an answer to the question why 'so few of the hours of life are filled with objects adequate to the mind of man', Johnson turns in *A Journey* not to the hours, as he had done earlier, but to the objects available to the human mind. In a description of the 'nakedness' of the area outside the village of Anoch,

Johnson holds that:

an eye accustomed to flowery pastures and waving harvests is astonished and repelled by this wide extent to hopeless sterility. The appearance is that of matter incapable of form or usefulness, dismissed by nature from her care and disinherited of her favours, left in its original elemental state, or quickened with only one sullen power of useless vegetation. (*Journey*, 39–40)

The important consideration here is the repudiation of the notion of a necessary and fruitful connection of 'matter' to 'form' and 'usefulness'. Deprived of this connection, matter is left to remain in its 'original elemental state', a rocky substratum to which vegetation can only be attached as 'decoration', a surface topping on a pre-existing and inert structure.

The attention Johnson devotes to this spectacle of 'hopeless sterility' may give rise to the uncomfortable feeling that his view of nature is inorganic. And so it is. For at bottom the ecological disequilibrium between stony matter and vegetative form is paralleled by a still more profound imbalance between natural scarcity and the insatiable human need for novelty. Thus even the 'flowery pastures' and 'waving harvests' will fail to satisfy this need: in a style characteristic of all of Johnson's descriptive writing, the phrase is deliberately generalized, lacking the sensuous particularity and richness of descriptions sustained by a vision of natural abundance and variety.

The point is not to dispute the accuracy of Johnson's account of the topography of the Western Islands. Rather, it is to observe the way this account is shaped in terms of an ontology of natural scarcity. From the perspective of this ontology, man is distinguished from the beasts by his efforts to raise himself above the level of mere subsistence: he is *Homo faber*, 'toolmaker'. As an exponent of this conception of man, Johnson seeks to suggest the ways individuals must act if they are to overcome the exigencies of a bleak and hostile environment. This is almost certainly what gives Johnson's *Journey* the polemical, didactic thrust that many readers have noticed; it is also why Johnson contends that the techniques of the manual arts are comparable in value to those of the liberal arts – the techniques of agriculture which make it possible for man to satisfy his needs for food, or those of cloth-making which make it possible for man to protect himself from the rigors of the weather. Johnson's emphasis on these craft techniques in *A Journey to the Western Islands* is not accidental; it is a necessary outgrowth of his vision of man and nature.

*   *   *

At the heart of Johnson's vision of man and nature is the assumption that human beings are inextricably related to the landscape they inhabit. The deprivation that Johnson encounters is not only present in a material form; it permeates the daily activities of the people and extends to their psychological attitudes, their social organization, and their manners. Thus the fragments and ruins that Johnson finds in his tours of monasteries and forts is paralleled by a fragmentation in the consciousness of the inhabitants, while the gradual and apparently irreversible deforestation of the Highlands is matched by the equally irreversible depopulation of its human communities. The expressed purpose behind Johnson's description of the two sets of events is not what is critical but rather the sensible resemblance between them. It is upon this sort of analogy that Johnson represents imaginatively to his readers the whole fabric of life in the Highlands and Western Islands. When he describes the barrenness of the land, he is not presenting an isolated fact. He is implying, in a concrete way, that the nudity of the fields is analogous to the silence and solitude of the streets. Considered point by point, the facts of Johnson's travel memoir appear simply arbitrary. 'History' has thrown them up, and history may efface or alter them. But seen as parts of an ordered whole, they become comprehensible, for they are linked to other parts by a system of implicit connections and dependencies.

A good example of the way an isolated fact becomes part of a larger complex in *A Journey to the Western Islands* can be seen in Johnson's account of the silence he finds permeating all aspects of Highland life. He discovers this silence in the cities and on the roads; in the highest and lowest levels of society. Thus it encompasses not only isolated individuals like 'the old woman' dwelling quietly in the vault of a religious building in St Andrews (*Journey*, 8–9), but also the society of beggars who, far from being 'importunate' and 'clamorous', 'solicit silently, or very modestly' (*Journey*, 12). It is only by recognizing the pervasiveness of this silence that we can understand how it defines a world in which individuals and families dwell in relative isolation from one another. The evocation of this silent world is Johnson's way of conveying his sense of a people who live in poverty and depend only on themselves for the bare necessities of their existence.

This vision of Highland culture as an imaginative totality does not, however, explain what has prevented the inhabitants from trying to alleviate their situation. In trying to account for their passivity, Johnson brings together the notion of deprivation and that of closure: the imagery of mountains and caves, castles, huts, and cellars, connotes the short views that prevent men from trying to improve their lot. Scarcity and self-sufficiency are here in ironic equilibrium: the mountains, unlike those of the prelapsarian Happy Valley, do not serve to establish a boundary between inside and outside, good and evil. These oppositions are erased by a perspective that places the individual outside the closed circle of perfection, in a post-lapsarian space of natural scarcity and historical decline. This space is exemplified early in the narrative by the two deserted towers of the monastery at Aberbrothick. Boswell, Johnson writes, 'scrambled in at a high window' of one tower, 'but found the stairs within broken, and could not reach the top' (*Journey*, 11). Of the other tower, Johnson reports that 'we did not immediately discern the entrance, and as the night was gathering upon us, thought proper to desist' (*Journey*, 11). The failure of the townspeople to restore these towers becomes the symbol of a short-sightedness comparable to the failure of the Scots to plant trees. In both cases, the narrowness of perspective corresponds to a phenomenological vantage point which is at once enclosed and close to the rocky ground. The inhabitants have achieved self-sufficiency but they fail to see out over things and thus remain in direct contact with harsh reality. Johnson is convinced of the deleterious effect of this perspective on their daily lives: 'they are not commonly dexterous: their narrowness of life confines them to a few operations, and they are accustomed to endure little wants more than to remove them' (*Journey*, 30).

If the majority of inhabitants seem confined spatially to a world of limited perspectives and confined views, it is hardly surprising to find that they are also cut off from the past and future, enclosed in a narrow, shadowy present. Discovering to his discomfort, that the windows in town houses are 'closely shut' and thus afford no air for ventilation, Johnson notes that 'the art of joining squares of glass with lead is little used in Scotland, and in some places is totally forgotten' (*Journey*, 21). The inability to transmit knowledge of such manual arts from one generation to the next is symptomatic of a broader amnesia: 'hopeless are all attempts to find any traces of Highland learning. Nor are their primitive customs and ancient manner of life otherwise than very faintly and uncertainly remembered by the present race' (*Journey*, 112). This collective lapse of memory is related to a similar failure to plan for the future. Johnson believes that 'few regions have been denuded like this, where many centuries must have passed in waste without the least thought of future supply' (*Journey*, 10). Such thoughts are difficult, however, for a people condemned by natural scarcity to live in the present moment, for 'plantation is naturally the employment of a mind unburdened with care, and vacant to futurity, saturated with present good, and at leisure to derive gratification from the prospect of posterity' (*Journal*, 139).

Johnson's argument turns (as one might expect) on the relation between the mind and the immediate data of consciousness. Essentially the difficulty lies in the fact that the mind must be 'saturated' by a present plenitude before it can be 'vacant' to the future. Elsewhere in his writings, Johnson makes it clear that the mind doesn't require much to reach this saturation point, but it is only at that point, nonetheless, that the reason – a technical and calculating faculty that possesses a power of anticipation exceeding the givens of the senses – will able to inaugurate the possibility inscribed in the notion of material progress. When the mind expends all of its energy trying to fill a present deficiency, the rational faculty will remain dormant, incapable of going beyond what is immediately perceived.

Thus there is no difficulty in explaining how this passivity is reflected in the deplorable level of material culture in the Highlands. When Johnson is told at Inverness, for example, 'that the people learned from Cromwell's soldiers to make shoes and plant kail', he asks:

> how they lived without kail, it is not easy to guess: they cultivated hardly any other plant for common tables, and when they had not kail they probably had nothing. The numbers that go barefoot are still sufficient to shew that shoes may be spared: they are not yet considered as necessaries of life; for tall boys not otherwise meanly dressed, run without them in the streets and in the islands; the sons of gentlemen pass several of their first years with naked feet.
>
> (*Journey*, 28)

The example of the 'sons of gentlemen' is particularly illuminating in this respect, since even this relatively privileged group are but part of the primitive cultural level that Johnson finds almost everywhere in the Highlands. At least some communities are completely unfamiliar with the construction of houses: gardening and agriculture are reduced to their most 'laborious' and rudimentary level; and the clans of several families, which are joined under the leadership of a laird who no longer possesses any real authority, seem to be motivated entirely by the requirements of food-gathering and the ever-present threat of famine.

This rudimentary level of existence puts a phenomenology of penury into practice. The basic element of this phenomenology, as we have seen, is its acceptance of the status quo. The natural scarcity of existence in the Highlands and Hebrides imposes upon its inhabitants an imperative that requires them to concentrate all their energies upon what lies immediately before them. There is apparently but one constant, the overriding desire for survival. From the total enclosure of such inhabitants within their own lives, it follows that they are so unable to imagine any other kind of existence that they are wholly without perspective on their own lives. It is impossible for them to see themselves as others see them, and they are thus able to endure such hardships that would shock or disgust others. As Johnson observes:

> he that pines with hunger, is in little care how others shall be fed. The poor man is seldom studious to make his grandson rich. It may soon be discovered, why in a place, which hardly supplies the cravings of necessity, there has been little attention to the delights of fancy, and why distant convenience is unregarded, where the thoughts are turned with incessant solicitude upon every possibility of immediate advantage.   (*Journey*, 139–40)

Johnson was clearly fascinated and repelled by the process of acclimatization whereby an individual or a group can become accustomed to an existence that would be intolerable to anyone else. Yet it is important to note that Johnson never commits himself wholly to a naturalistic explanation of this passivity. On the contrary, he is clearly ready to perceive it, not as a permanent state, but as a possible regression from an earlier, more developed level of culture.

Johnson initially finds evidence for this more developed level of culture in the ruins he encounters. The centres of Scottish learning appear as a jumble of decayed and crumbling forms, impressive walls and empty vaults. Their ruins encompass not only their monasteries, cathedrals, and forts, but also their ancient cities. St Andrews, for example, has 'gradually decayed'; of the cathedral, only 'the foundations may still be traced, and a small part of the wall is standing', while 'not far from the cathedral, on the margin of the water, stands a fragment of the castle in which the archbishop anciently resided'. At the time of Johnson's visit, 'one of its streets' was 'lost, and in those that remain, there is the silence and solitude of, inactive indigence and gloomy depopulation' (*Journey*, 6). St Andrews is perhaps the extreme case, but the monastery at Aberbrothick too is in a state of delapidation (*Journey*, 11). The cathedral at Elgin lies in 'ruins' (*Journey*, 23), while the fort at Inch Keith, which Boswell and Johnson traversed by climbing over 'shattered crags', also lies in 'ruins', though Johnson characteristically observes, 'not so injured by time but that it might easily be restored to its former state' (*Journey*, 3).[13]

These instances of fragments and ruins are not simply exploited for their pleasing, picturesque qualities. They are the visible manifestation of a process of degeneration that has led to what Johnson sees as the current state of stagnation in the liberal arts. This state has been the result of cataclysmic eruption (the Reformation) rather than gradual deterioration, but the end result would have been the same in either case – the destruction and levelling of an earlier, more complex culture. Such a process is only partially counterbalanced by the evidence of material progress that Johnson finds in Glasgow and the new town at Aberdeen. Even in the lowlands, the overall impression conveyed by Johnson's narrative is one of decline, not progress.

This is undoubtedly why Johnson does not use the barrenness of Scottish life as a pretext for a sentimental celebration of 'savage virtues and barbarous grandeur'. Johnson believes that the Highlands has also undergone a change, but that a change which produced conditions of life so elementary that they suggest a primitive state must have amounted to a regression. The existence of this state is thus tied in part to the conquest of the Highlands by the English at the battle of Culloden in 1746. The consequences of this event are so thoroughgoing that they virtually effaced the feudal system they overthrew: 'there was perhaps never any change of national manners so quick, so great, and so general, as that which has operated in the Highlands, by the last conquest, and the subsequent laws' (*Journey*, 57). Once the English traveller might have found a culture radically different from his own awaiting him at the end of his journey to the Westernd Islands.[14] Now, in his own personal *Tristes Tropiques*, Johnson finds only impoverished imitations of his own culture, set off here and there by relics of an almost forgotten past:

> we came thither too late to see what we expected, a people of peculiar appearance, and a system of antiquated life. The clans retain little now of their original character, their ferocity of temper is softened, their military ardour is extinguished, their dignity of independence is depressed, their contempt of government subdued, and their reverence for their chiefs abated. Of what they had before the late conquest of their country, there remain only their language and their poverty. Their language is attacked on every side. Schools are erected, in which English only is taught, and there were lately some who thought it reasonable to refuse them a version of the holy scriptures, that they might have no monument of their mother tongue. (*Journey*, 57–8).

Johnson's discovery of this transformation of the different into the same almost appears to alter his travel-memoir in mid-course. Instead of concentrating all his efforts upon the careful study of an alien culture, he must turn part of his attention to the reconstruction of a vanished way of life. What complicates this task, however, is that the main aspects of this way of life can only be given, not through direct observation but through inference and interrogation. The result is an account that is neither exactly speculative theorizing nor empirical anthropology but an interweaving of both.

* * *

A significant part of Johnson's attempted recreation of the vanished life of the Highlands is devoted to the institutions of feudalism. Much of what Johnson was able to discover obviously attracted him – the system of insular subordination and ranking, the power and authority of the lairds, their direct engagement in agriculture. He fully appreciated the impact of commerce upon Highland feudalism and expressed the wish that 'no change' in the ancient practice of 'payments of rent in kind' had been made:

> money confounds subordination, by overpowering the distinctions of rank and birth, and weakens authority by supplying power of resistance, or expedients for escape. The feudal system is formed for a nation employed in agriculture, and has never long kept its hold where gold and silver have become common. (*Journey*, 113)

Such sympathies might appear to place Johnson in apparent conflict with English mercantilism, with its commitments to trade, comfort, money, and the rapid circulation of goods and services. But Johnson believed that these innovations were inevitable, even desirable. How, then, can Johnson's evident nostalgia for a vanished feudalism be reconciled with his evident admiration for the benefits wrought by commerce?[15] In the hands of his predecessor, Thomas Pennant, this conflict never appears, and the evaluation of economic growth emphasizes its material aspects without any concern as to untoward moral consequences.[16] But in Johnson's hands, the interplay of material progress and moral decline takes the form of a subtle dialectic that invests *A Journey* with remarkable depths. One side of this dialectic can be seen in Johnson's refusal to portray Scottish feudalism as a self-contained, self-authorizing source of legitimacy. Its social and political ranks are a reflection, not of the hierarchy of creation, but of a universal jostling for either security or pre-eminence. That the ordering of men into these ranks must at some stage have required coercion rather than consent is evidenced by the violence and endemic warfare with which the system of feudal subordination is seen to be inextricably linked.

Johnson's conception of a violent hierarchy is not confined to Highland feudalism but is viewed as a typical characteristic of all feudal societies. To be sure, Johnson expresses a strong admiration early in the Vinerian lectures for the distinctions and privileges which are a characteristic feature of feudalism:

> no constitution of which history gives us any intelligence produced such regular and natural subordination as the feudal system, which ranged a people in successive gradations, from the king or general proprietor, till a long train of tenures and subinfeudations ended in the churl who gleaned his bread from the land of another.
>
> (*Vinerian Lectures*, 103)

Yet the intricate system of distinctions described in this passage is only the outgrowth of a long process of development. Indeed, the character of early feudalism, as Johnson perceives it, was so violent that the clergy were compelled to become the chief preservers of culture:

> In those times of ferocious barbarity which succeeded the first establishment of feudal power, when there was yet no regular distribution of civil justice, when the laws (imperfect as they were) were unknown to the greatest part of the people and, for want of a gradual subordination of power, were violated by every chieftain who was stronger than his opponent, it was natural for the oppressed and timorous to fly for shelter to the ministers of religion, whose character restrained them at least from open violence and *avowed* contempt of reason and of justice.   (*Vinerian Lectures*, 110−11)

Although this climate of fear is most prominent in early feudalism, it is never entirely effaced from the 'feudal system'. This is partly because fear of the neighbour who would devour serves in Johnson's political thought as the central motive in the formation of all societies. It is the desire for self-preservation rather than domination which encourages men to band together, an assumption that Johnson explains by reference to the truth that 'the dread of evil may be always made more powerful than the appetite of good' (*Vinerian Lectures*, 110). This means that there is no pre-existing social organization to which all members of a community naturally belong. They are rather envisaged as isolated individuals, existing without any social ties, who 'fly for shelter' into feudal organizations because of their need for personal security. Johnson, accordingly, envisages the main function of 'money' as a partial reversal of this process: money acts to constrict political authority, loosening ties and allowing individuals to 'escape' from traditional forms of legally enforceable subservience.

Thus the actual shift described in *A Journey to the Western Islands* is not from natural unity to artificial division. If such a shift ever occurred, it must have taken place before the formation of society itself, for feudalism is already subject to difference and division. Hence it is not surprising that the theoretical model Johnson employs to represent the 'system of antiquated life of the Highlanders' to his readers is, at least one critic has observed, perhaps best described as Hobbesian, inasmuch as it presupposed a generalized state of discord rather than harmony.[17] Like Hobbes's *omnium bellum contra omnes*, Johnson's feudalism emerges in an atmosphere of fear and violence that seems to give rise to an unending cycle of rapine and revenge. It is true that Johnson differs from Hobbes in locating this universal war, not in an ahistorical state of nature but in a historical framework that extends from an early period of 'ferocious barbarism' to the late feudal world of the fifteenth, sixteenth, and seventeenth centuries. In the latter era, men are united into warring groups, subsisting by farming, gardening, and herding. Yet even this disparity may be less than it appears, for Johnson's earliest period – an era of nomadic hunter–gatherers (*Journey*, 37, 72) – is as much a theoretical construct as the Hobbesian state of nature.

According to Johnson's Hobbesian schema, the passion of fear dictates the primary architectural form of feudal life – the fortress:

as the inhabitants of the Hebrides lived, for many ages, in continual expectation of hostilities, the chief of every clan resided in a fortress. . . . They had formerly reason to be afraid, not only of declared wars and authorized invaders, or of roving pirates, which, in the northern areas, must have been very common; but of inroads and insults from rival clans, who, in the plenitude of feudal independence, asked no leave of their Sovereign to make war on one another. (*Journey*, 68)

The animal enclosure is another structure appropriate to an age dominated by the politics of fear and violence:

in lawless times, when the inhabitants of every mountain stole the

cattle of their neighbor, these inclosures were used to secure the herds
and flocks in the night.   (*Journey*, 72)

The mentality embodied in the fortress and animal enclosure does not,
however, give rise to a defensive polarity of inside vs. outside, as it does
in *Rasselas*, for this mentality is only the reverse side of a posture of
hostility and aggression, a posture that lasted until the Highlanders were
finally disarmed:

> Till the Highlanders lost their ferocity, with their arms, they suffered
> from each other all that malignity could dictate, or precipitance could
> act. Every provocation was revenged with blood, and no man that
> ventured into numerous company, by whatever occasion brought
> together, was sure of returning without a wound.   (*Journey*, 92)

Johnson's emphasis on fear and violence thus irresistibly suggests the
ideas of Hobbes. There is no question of a direct influence, of course, but
rather of a similarity of perspective. Indeed, Johnson differs from Hobbes
in one major respect. Although Gordon Schochet has shown that the
Hobbesian model was flexible enough to include patriarchal families and
small groups, its basic orientation remains individualist – it is defined as a
war of every man against every man.[18] Highland feudalism, by contrast, is
defined in terms of a congerie of clans led by lairds. As the 'original'
owners of the 'land', the lairds possess a power so great that excels even
that of kings: 'kings can, for the most part, only exalt or degrade. The laird
at pleasure can feed or starve, can give bread, or withhold it' (*Journey*, 85).
Whatever degree of support prevailed depended, however, not only on
force or on an original tenure, but also on 'the kindness of consanguinity
and the reverence of patriarchal authority' (*Journey*, 85). Legal power and
patriarchalist legitimacy were thus eventually fused:

> every duty, moral or political, was absorbed in affection and adherence
> to the chief. Not many years have passed since the clans knew no law
> but the laird's will. He told them to whom they should be friends or
> enemies, what King they should obey, and what religion they should
> profess.   (*Journey*, 86)

However, the very absolutism of the laird's power makes it possible for
Johnson to fit it into his Hobbesian conceptual model. In the absence of a
centralized authority, the relation between clans has gravitated into the
kind of lawless warfare that can be rendered in generalized, schematic

terms. And the laird's absolutism ensured that this warfare would extend downward, from the apex to the very lowest orders of the clan. It may be for this reason that Johnson employs the notion of the social compact to describe the disarming of the clans. To be sure, Johnson shows a healthy scepticism concerning the efficacy of the compact. Its effectiveness depends for Johnson, as it does for Hobbes or Locke, upon its power to afford protection for a nation's subjects:

> the supreme power in every community has the right of debarring every individual, and every subordinate society from self-defence, only because the supreme power is able to defend them: and therefore where the governor cannot act, he must trust the subject to act for himself. (*Journey*, 90–1)

The very distance of the sovereign power from the subjects appears in this case to militate against its capacity to give the kind of protection the exchange of freedom for security requires:

> these islands might be wasted with fire and sword before their sovereign would know their distress. . . . Laws that place the subjects in such a state, contravene the first principles of the compact of authority: they exact obedience and yield no protection. (*Journey*, 91)

This contrast between the ideal of protection and what can be achieved in practice corresponds to Imlac's scepticism about the capacity of governors to govern in *Rasselas*. Yet the fact that Johnson evinces considerable doubt about the viability of the compact in no way mitigates against its validity as a conceptual model. The underlying aim of such a model is to provide a framework against which an unfamiliar experience can be rendered comprehensible to the contemporary reader.

Whatever Johnson's veiw of the compact theory, its basic premise, like that of the universal war, is derived from the writings of a Hobbes or a Locke rather than from the eulogists of the great chain of being. In spite of his nostalgia for the feudal system of subordination, Johnson views it in relation to the lawlessness of the Highlanders. The reasons for this lawlessness are not simply cultural, moreover; they are also natural. Johnson shared the view of Montesquieu and Lord Kames that the spirit of a people is not just determined by their form of government but also by extrinsic causes like nature and climate. But Johnson could not remain content, like Montesquieu or Kames, with general assertions about the influence of cold vs. hot climates. His *Journey* rather resembles Fernand Braudel's *The Mediterreanean and the Mediterranean World in the Age of Phillip II* in its emphasis on the unity of specific geographical and historical factors. Indeed, Braudel argues explicitly for what seems to be an assumption of Johnson's *Journey*: the close relationship between the ecological characteristics of mountain regions of the Highlands and the ferocity, isolation and independence of its inhabitants.[19] Johnson reports that 'mountainous countries commonly contain the original, at least the old race of inhabitants, for they are not easily conquered' (*Journey*, 43). The mountain habitat is unsuitable not only for agriculture but also for hunting and gathering, since vegetation and animal life are necessarily scanty. The lowlands, by contrast, are potentially prodigal in wildlife, and the soil is fertile so long as it is properly cultivated. The mountainous terrain thus exerts a negative influence in Johnson's analysis; it does not open up new possibilities but, rather, restricts those of the lowlands:

> the inhabitants of mountains form distinct races, and are careful to preserve their genealogies. Men in a small district necessarily mingle blood by intermarriages, and combine at last into one family, with a common interest in the honour and disgrace of every individual.
>
> (*Journey*, 47)

Even more striking is the isolated mode of existence which the mountainous terrain necessarily imposes upon its inhabitants:

> as mountains are long before they are conquered, they are likewise long before they are civilized. Men are softened by intercourse mutually profitable, and instructed by comparing their own notions with those of others.
>
> (*Journey*, 43–4)

This conception of the interrelation between nature and culture, so prominent in Johnson's analysis of feudalism, may help to account for an unusual aspect of *A Journey to the Western Islands*. This is his view of what he once called 'heroic romance'. Rather than regarding heroic romance as the product of a 'wild strain of imagination' that is divorced from daily life, as he did in *Rambler*, No. 4, Johnson now believes that it is inseparable from the culture that produced it. Indeed, Johnson is at pains to show how the underlying structure of feudalism is strengthened and perpetuated through the imaginative expression of the social values upon which it rests:

These castles afford another evidence that the fictions of romantick chivalry had for their basis the real manners of feudal times, when every lord of a seignory lived in his hold lawless and unaccountable, with all the licentiousness and insolence of uncontested superiority and unprincipled power.   (*Journey*, 155)

Here, tales of romantic chivalry are seen as having their 'basis' in what Johnson sees as the most characteristic feature of Highland feudalism – its emphasis upon power. To illustrate how deeply this emphasis is ingrained in the world-view of the Highlanders, Johnson relates several popular stories in which an underlying struggle for power leads to treachery and revenge.[20] These stories provide models of action that highlight for the English reader the conflicts which the inherent instability of Highland culture inevitably generated. When this instability is transposed from the cultural to the natural realm, it is transformed from a conflict between man and man to a conflict between man and mythic beast. Johnson describes how this conflict is, in turn, but part of a larger conflict between nature and culture, forest and castle, solitary gloom and communal festivity:

> the fictions of the Gothick romances were not so remote from credibility as they are now thought. In the full prevalence of the feudal institution, when violence desolated the world, and every baron lived in a fortress, forests and castles were regularly succeeded by each other, and the adventurer might very suddenly pass from the gloom of woods, or the ruggedness of moors, to seats of plenty, gaiety, and magnificence. Whatever is imaged in the wildest tale, if giants, dragons, and enchantment be expected, would be felt by him, who, wandering in the mountains without a guide, or upon the sea, without a pilot, should be carried amidst his terror and uncertainty to the hospitality and elegance of Raasay or Dunvegan.   (*Journey*, 77)

In affirming the relationship between 'Gothick romances' and the social system from which they emerged, Johnson is referring not only to their themes but also to their origins and mode of composition. Johnson's conviction that *Ossian* is a forgery is not based upon arguments about individual versus multiple authorship, or oral versus literary composition, as we might suppose. It rather stems from his belief in the violence and discontinuity of Highland life. This belief leads naturally to the view that 'in an unwritten speech, nothing that is not very short is transmitted from one generation to another' (*Journey*, 116). Theories of epic improvisation and formulaic texts thus become irrelevant to written texts, which are not to be found at the end of a long chain of hypothetical oral versions, but stand, like the Homeric epics, at the beginning of a literary or written tradition. This belief helps Johnson dismiss the existence of an antique tradition of Highlander epic and romance. The author of *Ossian* is identified as Macpherson and its provenance is shown to be of the eighteenth century.

There is thus a profound difference for Johnson between oral and literary composition. Oral composition is ephemeral, that is, it cannot transcend its own limitations. Literary culture is permanent and contains within itself the potentiality for transmission and circulation. But the permanence of literary works also derives from another source. The system is not self-sufficient, and its frame of reference is always the surrounding culture. The opposition between oral and literary composition thus also implies the opposition between feudalism and mercantilism. Where feudalism is defined in terms of vertical hierarchies whose internal conflicts inhibit true exchange, mercantilism is a horizontal network in which a strong central authority makes possible the easy circulation of money, books, material goods and services.

It is in this way that feudalism and mercantilism come to be opposed in *A Journey to the Western Islands*. With his characteristic sense of realism, Johnson believes that commerce is superior to feudalism in many respects, yet he is acutely aware of its deficiencies. For Johnson, its values are disruptive of the sense of patriarchal loyalty and obligation that are at the heart of clan life:

> Since the islanders, no longer content to live, have learned the desire of growing rich, an ancient dependent is in danger of giving way to a higher bidder, at the expense of domestick dignity and hereditary power. The stranger whose money buys him preference, considers himself as paying for all that he has and is indifferent about the laird's honour or safety. The commodiousness of money is indeed great; but there are some advantages which money cannot buy, and which therefore no wise man will by the love of money be tempted to forgo. *(Journey, 86)*

Yet the sense of loyalty already contains within itself the premise of its anithesis — 'the desire for growing rich'. For the existence of peace and a strong central authority is accompanied by a psychological transformation of the fear that is always a part of feudal obligation to the desire for material reward. The social transformation that Johnson describes as altering Scottish values is thus, in one sense, the result of the introduction of money and commerce. Yet, in another sense, it is only a metamorphosis of the intrinsically competitive and agonistic strain that Johnson finds in Scottish feudalism and indeed in all cultures. Johnson never explicitly considers such a metamorphosis in the *Journey*, yet he assumes throughout the text that this strain is as likely to disrupt as to reinforce social ties, to engender strife as to enforce harmony. Thus the displacement of Highland feudalism by the beginnings of mercantilism does not introduce a wholly new source of conflict because feudalism is itself the embodiment of conflict.

*   *   *

What we find in feudalism, then, is an element corresponding to the desire for material gain in commercial nations. Yet this element – the predominance of some form of feuding and violence between clans – was not by itself sufficient, in Johnson's view, to provide an impetus for historical change. Highland feudalism was not spared sources of instability, but it was nonetheless relatively stable internally, a system of warring groups that maintained their cohesion through opposition. Highland feudalism benefited, moreover, from the gradual merging of power and authority, the real and the ideal, that Johnson sees as a characteristic feature in the development of all early nations. In such nations, ceremonies play a major role in establishing the customs, the sustaining fictions, by which arbitrary power and violence come to be legitimized:

> It will be found by observation that when countries emerge from barbarity, the first approaches to elegance and civility are made by public celebrations and formal magnificence. While men are gross and ignorant, unskilled in the arts of reasoning, and enjoying few opportunities of information and instruction, they are much influenced by unusual and splendid ceremonies which strongly affect the imagination and leave deep impressions on the memory.
>
> (*Vinerian Lectures*, 93)[21]

Because of the power of such sustaining fictions, feudalism is seen as having become entrenched in the Highlands; in spite of its violence, there is no internal principle by which a smooth transition from feudalism to mercantilism could be produced. Late feudalism was indeed the end product of an evolutionary movement from lawlessness to stability, but the passage from one system of exchange to another – from feudalism to commerce – cannot be understood as the result of any such movement; it can come about only through a rupture, a catastrophe:

> established custom is not easily broken, till some great event shakes the whole system of things, and life seems to recommence upon new principles.   (*Journey*, 10)

Nothing within feudalism itself could produce such a rupture or reason for leaving it. The causality for the catastrophe had to be natural and exterior to it.

The model for this principle of historic causality appears to be natural and physico-anthropological. In *Metahistory*, Hayden White contends that what he calls the 'Epic form' of eighteenth-century historiography:

> presupposes the cosmology represented in the philosophy of Leibniz, with its doctrine of continuity as its informing ontological principle, its belief in analogical reasoning as an epistemological principle; and its notion that all changes are nothing by transformations by degrees from one state or condition to another of a 'nature' whose essence changes not at all.[22]

Johnson's conception of historical change, however, is Newtonian rather than Leibnizian in White's sense, for it is not a function of the principle of continuity or a manifestation of analogical reasoning. In contrast to the Leibnizian notion of an internal transformation by degrees, change for Johnson is external and disruptive. Thus the shattering impact of Knox and the Reformation on the medieval Scottish church, or the tragic consequences of the Conquest of 1746 on Highland feudalism resemble nothing so much as the devastation wrought by the Black Spring of '71 on the lives of the people of the Western Islands (*Journey*, 78, 137). This does not mean that the potentiality for drastic change is not inherent in all societies. But what prevents this potentiality from being realized is the natural inertia of 'established custom'. This inertia is not just one empirical characteristic among others. It is a determining feature of what Johnson calls 'the system of things' and thus corresponds to the psychological inertia of temporal continuity. Within the Newtonian schema, the *vis inertiae* is natural. Since change cannot be borne out of inertia, it must be produced through the impact of an exterior force or body. Once the impact occurs in a culture, moreover, the system of life is not easily restored. The long-term consequence of the catastrophe is not necessarily the establishment of a new system but rather a stagnation that is unsustained by custom or tradition.

This catastrophist theory of historical change may very well influence Johnson's assessment of the impact of the English conquest upon Highland life. He tries to give a balanced picture of this impact. Every change has 'its evil and its good' in Johnson's view, yet when the two are compared, it seems evident that the evil outweighs the good in Johnson's account. Radical change has destroyed Scottish feudalism, but it has left in its wake only a void. If the 'plenitude' of Highland feudalism was mirrored in the authority of its 'patriarchal rulers', so the fragmented and decaying culture of the late eighteenth century is reflected in the transformation of these same rulers into 'rapacious landlords' (*Journey*, 89). For Johnson, the consequences of such a transformation are embodied less in material than in psychic terms. The cultural demoralization of the Scottish people can be seen in its impact upon their sense of self-esteem:

> their pride has been crushed by the heavy hand of a vindictive conqueror, whose severities have been followed by laws, which, though they cannot be called cruel, have produced much discontent, because they operate on the surface of life, and make every eye bear witness to subjection. To be compelled to a new dress has always been found painful. (*Journey*, 89)

The positive factors – the lessening of ignorance, the diminution of violence, and the melioration of poverty – seem inadequate to compensate for the debilitating effects of this loss of dignity upon the populace as a whole.

Johnson's realism is reflected in his effort to preserve a measure of detachment toward the Conquest and its legacy. He achieves this detachment, not by remaining aloof from controversy, but, as in *Rasselas*, by alternating between the two sides. He appears to believe in both feudalism and commerce – in subordination and liberty – and to disbelieve in them both as well.[23] More than this, he becomes so preoccupied with the problem of documentation that it almost seems to overshadow the political issue. Indeed, the silence, passivity, and general negligence which are so important in Johnson's account of his difficulties in gathering evidence are themselves seen as the inevitable result of the interposition of an alien body of laws from the outside. A system of values has been destroyed, but a new system has not as yet taken root. Inasmuch as a culture can only be understood through a knowledge of the principles by which its system of life is governed, the result of the destruction of that system is uncertainty. Johnson pointedly draws attention to this uncertainty in his account of domestic servants:

> the condition of domestick servants, or the price of occasional labour, I do not know with certainty. I was told that the maids have sheep, and are allowed to spin for their own clothing; perhaps they have no pecuniary wages, or none but in very wealthy families. The state of life, which has hitherto been purely pastoral, begins now to be a little variegated with commerce; but novelties enter by degrees, and till one mode has fully prevailed over the other, no settled notion can be formed. (*Journey*, 89)

The confusion of class roles perceived by Johnson in a culture where 'the maids have sheep, and are allowed to spin for their own clothing' is symptomatic of his epistemological and semiotic difficulties. As a traveller, Johnson seems condemned to visit such gracious hosts as he can understand precisely because his own culture has already transformed them into simulacra of himself, or to dwell among those who, not so transformed, are for that reason largely 'illiterate' and thus wholly absorbed in the struggle for survival. Either he is a wanderer among true savages who display 'neither shame from ignorance, nor pride in knowledge; neither curiosity to inquire, nor vanity to communicate' (*Journey*, 111), or he is dependent upon the unwritten history preserved by the chiefs. Confronted with the fact that 'the chiefs were sometimes ignorant and careless, and sometimes kept busy by turbulence and contention' (*Journey*, 111), Johnson finds the oral tradition by which their exploits are transmitted unreliable as a source of data. Subject to the same discontinuities as the culture which produced it, oral tradition is 'but a meteor, which, if once it falls, cannot be rekindled' (*Journey*, 111).

Johnson's scepticism concerning sources is thus the direct result of a conception of conflict and historical causality which appears to make impossible any notion of transition by infinitesimal degrees. His determination to gather data in the face of such an obstacle is a corollary of his passionate desire to know: to satisfy that desire, any reconstruction of the past must be supported by 'facts'. This suggests two of the features of modern anthropology that Johnson anticipates: its abstention from speculation about the past and its rigorous, even guilt-ridden standards of documentation.[24] Yet Johnson goes even further, subverting the very assumption of a stable, coherent alterity behind the vicissitudes of history: 'what memorials were to be expected from an illiterate people, whose whole time is a series of distress' (*Journey*, 110). Yet this scepticism in turns poses a further problem: although there are no historical discoveries, the desire for knowledge is insatiable. What remains is an ethnography of surfaces, an empire of signs without an essential content.

In spite of Johnson's sense of ultimate failure, however, *A Journey to the Western Islands* remains one of his most impressive works – a major attempt at a new kind of narrative and meeting ground for projects felt to be contradictory. On the one hand, it displays a powerful ethnographic impulse, examining a culture in terms of its most prominent features, identifying and organizing them in a rationalistic spirit. In its attempt to distinguish what is 'peculiar' to Scottish culture, it contributes to the programme outlined in *Idler*, No. 97. Yet *A Journey to the Western Islands* also reveals the limitations of that project. Johnson discloses that instead of treating Scottish culture as the manifestation of an underlying system, he is forced to explore its differences with itself; its unrecognizable alterity; and the way it has degenerated from a past which is only dimly recognizable. The history of Scotland is only partially accessible to us, in Johnson's opinion, and the tension between what is known and what, because of the paucity of oral traditions, remains forever beyond our reach, constitutes part of the greatness of his experimental narrative.

8
Conclusion

Johnson's aim in his review of Jenyns's *Inquiry* was to destroy the notion that the cosmos has to be explained in terms of the great chain of being. The very notion of plenitude, hierarchy, and continuity are dismantled; they give way to the notions of ontological scarcity, generic discontinuity, and temporal succession. Yet this does not mean, as some of Johnson's interpreters have concluded, that his conception of the spatiotemporal order of things is limited to that of the straightfoward empiricist. There is much more involved in a perspective that takes issue with the emphasis Jenyns places upon the concept of infinity in his explanation of the scale of being. This perspective has conceptual underpinnings that can be traced to currents of thought in Johnson's own time. It presupposes, I have argued, not only a Lockean empiricism but a post-Newtonian ontology in which the universe is viewed as a vacuum rather than a plenum.

It is not hard to show how Newton's ideas make contact with Johnson's thinking about Jenyns's metaphysics. Newton was squarely opposed, like Johnson, to any mode of reasoning which made *a priori* hypotheses the foundation of philosophical speculation. The preference for an approach that seeks to reveal the contradictions and gaps of meaning by which metaphysical certainties dissolve into their opposites is likewise common to Newton and Johnson. As Johnson demonstrates, Jenyns's argument is concerned not so much with the literal sense of philosophical terms, as with the 'uncertain and figurative' sense these terms acquire 'when they are applied to the works of Omnipotence'. The concept of infinity is inhabited by structures of implication which nowhere coincide with common usage. Infinity is in this sense a synecdoche for all those reversals and enlargements of meaning which occur whenever speculation overreaches itself. Such speculation threatens theological disquisition by drawing attention to the limitation inherent in *a priori* thought in general, the extent to which infinitude disrupts the illusion of a perfectly closed and coherent system. Johnson's procedure thus acts to suspend traditional conceptions of philosophical theology. It operates instead with logical paradoxes which bring out the radical disjunction between intention and sense, finitude and infinitude, what language explicitly says and what its figural workings constrain it to mean.

It may seem perverse to press too hard on what is after all only a relatively late review in a massive and diverse body of writing. And indeed it would be if there were not a number of complicating factors. Johnson was clearly exposed to the new post-Newtonian current of ideas long before he undertook to review Jenyns's treatise. Moreover, there is the well-known fact that the theme of vacuity occurs, in a great variety of guises, throughout his writings. Thirdly, it seems evident that Johnson wrote what amounted to a Newtonian critique of Jenyns's theodicy. The outcome of this critique – so far as it has one – comes down to an interstitial vacuum (the vacuities of Newton are Johnson's most frequent recourse in exposing the logical fallacies in Jenyns's philosophical argument).

Johnson's rejection of philosophy as a systematic edifice, in favour of a much more scaled-down perspective, was of course nothing new in the history of English thought. Bacon and Locke, as well as Newton, had reacted against rationalistic metaphysics for much the same reason, though to very different ends. For Locke, this meant an epistemological alternative to the cob-webs spun by those who use words without any clear and distinct idea of what they mean. For Johnson, on the contrary, it led to an insistence that ethics, not metaphysics or epistemology, was the proper study of mankind. Johnson is a moralist, conscious of his affinities with Addison or Burton's *Anatomy of Melancholy*. The notion has a literary, even journalistic stress that has no immediate analogy with Locke or Berkeley. The moralist can turn to drama, journalism, or fiction like Addison or Fielding. Or he may, like Johnson, also work outward from what is, in their origin, highly specialized and diverse fields of interest like lexicography, criticism, and travel writing.

Only the linguist may be equipped to pass judgment on the technical solutions that Johnson put forward to the complex problems of English language and lexicography. But the issues that Johnson raised in his critique of Jenyns's *Inquiry* have a direct bearing on the broader implications of his moral outlook. In his essays in the *Rambler*, *Idler*, and *Adventurer*, Johnson sets out to question an ethics of plenitude that for him is based on what he believes to be an interior rhetoric of self-deception. This rhetoric holds out the prospect of a perfect coincidence between desire and fulfilment while achieving that end by what amounts to illusory means. Johnson relentlessly pursues the evasions and subterfuges by which a desire for plenitude leads man to reduce heterogeneity to unity, temporal succession to an eternal moment of undifferentiated happiness. Acquiescence in this desire becomes, for Johnson, virtually a species of madness. There may be 'no human mind' which, strictly speaking is 'in its right state', but the mind is not thereby absolved of the obligation of seeing through the ruses of its own too-willing deception.

Johnson reserves his fiercest polemics, however, for what he sees as the tactics of political evasion promoted by those who, like Jenyns, seek to invest the social hierarchy with the aura of sacred authority. Theirs is a form of sophistry, bent on concealing the fact that no political institution is exempt from the pressures and dislocations produced by the 'lust of dominion'. Political realism and sympathy for the oppressed alike require a repudiation of the view that the established order can be made explicable in terms of divine institution or natural analogy. Instead, Johnson draws attention, time and again, to the schism between the strong and the weak, the eminent and the obscure, thus emphasizing that any system of subordination is necessarily entangled in a universal contest for superiority. The politics of this struggle is complicated, however, by the fact that the *libido dominendi* is mediated by a uniquely human faculty, consciousness. Both man's imagination and his capacity for reflection and comparison result in the erection of a second illusory world of power relationships. The struggle for dominion thus remerges as a process in which the strong are ultimately and ironically dependent on the weak for their sense of self-worth. And the representation of this struggle depicts a process in which the original relation becomes subject to differing permutations and transformations. Thus the contradiction between the claims of self-sufficiency and dependency, of love and ambition in *Irene*, the dispossession of patriotic virtue by what is alien and parasitic in *London*, and the revelation that the warrior—conqueror is the slave of the slave in *The Vanity of Human Wishes* are three instances of the way in which the desire for mastery and self-identity is thwarted.

The *Plan* and *Preface* to *The Dictionary* can be read as an extension of the presuppositions of the universal struggle for dominion into the realm of language. A deluded overreaching underlies the disruptive tendencies of both the *libido dominendi* and the desire for linguistic innovation. There is an implicit parallel in Johnson's thought between the drive for self-sufficiency and the 'caprice' that generates the seemingly endless play of linguistic variation. Just as the former renders the notion of a permanent, stable hierarchy fatuous, so the latter calls into question the ideal of a language that is freed from the flux of historical change. Both social and linguistic institutions, for Johnson, are estranged from any ideal of permanence: the lexicographer can never expect to attain the internal consistency of general grammar, just as the political theorist can never pin his hopes on the continuity and coherence of the great chain of being.

I may of course by stretching the implications of Johnson's thought in the *Plan* and *Preface*; here I merely want to emphasize that Johnson sees language only as a heightened and dramatic instance of the problems besetting all human activity. Johnson's lexicography is a rigorous and self-denying discipline, always on the verge of denying its own provisional insights. That it takes such a cautious form is hardly surprising given Johnson's sense of the labyrinthine difficulties opened up by the rift between words and things. Once the lexicographer despairs — as despair he must — of translating the ambiguities of natural language into a register of clear-cut referential implication, then he has no other option but to try to master and minimize the endless complexities of linguistic variation. Yet this can only lead to a feat of interpretive reversal: quotations from established authors must be allowed to comment on the meanings of the words they were originally intended to illustrate. But this juxtaposition of definitions and illustrative quotations is only that, a juxtaposition. The means by which instances of recorded (often literary) usage can be brought to bear on definitions remains elusive. And this suggests in turn that language may be subject to a generalized arbitrariness, the effects of which are most clearly visible in the way meanings resist any easy identificaction with the graphic and phonetic qualities of words.

Johnson's approach to lexicography turns, then, on an attempt to contain and articulate the endless disseminating energies of language. But when Johnson turns from language to nature, from words to things in *A Journey to the Western Islands*, he finds a scarcity that contrasts with the illusory plenitude of man's linguistic invention. Regardless of whether this scarcity is further evidence for Johnson of the interstitial vacuities discovered by Newton and his followers, it shapes a cultural anthropology that gives force and direction to his exploration of the Highlands and Western Islands. His anthropological vision oscillates between a didactic and ironic apprehension of passivity and negligence and the tragic apprehension of a cycle of hybris and oppression that has left the Highlanders stranded in a cultural vacuum. This two-fold movement distinguishes the Whiggish from the conservative traveller, even though both perspectives take place in the shadow of a natural world that is harsh and barren. The imagery of emptiness, ruins, and physical catastrophe runs through *A Journey* like an obsessive motif. It is linked to Johnson's catastrophist idea of history as a series of losses and disasters, unredeemable except by a gradual amelioration of man's lot. But

Johnson insists that this amelioration can take place through foresight, through action, and through heroic exertion. Such an exertion is problematical in the sense that it carries with it the possibility of failure. But if individuals are unremitting in their efforts, they can contribute to a modest improvement in their level of material culture and comfort. And their experience of this comfort provides the basis for a limited transcendence of the limitations that penury and hardship inevitably place on humanity.

Yet the struggle between man and nature, like the struggle between man and man, is mediated through human consciousness. The incessant hunger of the imagination gives rise to a second illusory technology of total domination and unlimited transcendence. *Rasselas* portrays a world in which the habit of defining man's relation to the world in terms of the categories of survival and subsistence has been supplanted by different ways of conceptualizing it in terms of the categories of inside and outside, peace and war, solitude and society. Thus human technology can be comprehended in terms of the difference between an original, natural way of apprehending the world and various artificial alternatives. *Rasselas* depicts this difference as a contrast between an earth-bound realm of sober reasoning and craft technology and an arid sky-world of oratorical eloquence and mechanical marvels.

There is, however, an intrinsic irony in this transformation, for the cultivation of either pole of a binary opposition is invariably undermined by the other. The conventions of rhetoric contaminate logic, memories of society pervade solitude, and martial activity proves compatible with a pastoral mode of existence, so that man is forced into questions of choice which neither deny nor wholly affirm either alternative. The effect is to situate *Rasselas* on a shifting ground that oscillates between polarities without ever achieving the rhetorical limit of presence, authority, or closure.

But Johnson's conception of human destiny is based on more than the element of undecidability which prevents any clear-cut decision between issues of pleasure and pain, happiness and misery on the one hand or questions of good and evil on the other. It is also depends on his adopting this sceptical attitude only up to a point, in order to dispel narcissistic delusions that are likely to attend any choice. One can find many passages in *Rasselas* and elsewhere which call into doubt the claims of any ideal that holds forth the promise of an indissoluble fusion of mind and object, hope and fulfilment. Such are Imlac's remarks on 'that hunger of imagination that preys incessantly upon life, and must be always appeased by some employment'. For Johnson, however, this work of demystification is always at the service of a higher ethical or religious imperative. Johnson is always careful to insist that at some point there must be a redemption, however long postponed.

These are higher matters, however, on the uncertain boundary of what Johnson regarded as lying beyond the realm of human discourse. Implicitly they mock but are also vulnerable to the same self-critical reflections that characterize Johnson's probing of human delusion. The religious, however, is not the sole note in Johnson's register. A robust, earthy immediacy and common sense also quickened his meditations. The moments of negation that are repeated throughout his moral writings can be read as qualifications rather than cancellations of the moments of delusory exaltation that they accompany. There were of course darker moments, most obviously those in which Johnson gave expression to his horror of the void. But these moments are never allowed to dominate his writings. 'The reader' will find in his major texts 'no regions cursed with irremediable barrenness, or blessed with spontaneous fecundity, no perpetual gloom, or unceasing sunshine'. What these texts provide is a mode of discourse which combines an extreme ethical rigour with a wayward, uncertain movement that shuttles back and forth between vacuity and plenitude, so that neither is fully realized but neither can be negated: what Johnson describes as 'the causes of good and evil' move endlessly in the virtual space between pure negation and pure presence, between human misery and human happiness.

Notes

CHAPTER 1:   JOHNSON AND NEWTON

1.   On the question of Johnson's general interest in the New Science, I refer particularly to Richard B. Schwartz's richly informative *Samuel Johnson and the New Science* (Madison: University of Wisconsin Press, 1971). The manner in which Johnson adapted the terminology of the New Science in his writing has been effectively described by W. K. Wimsatt, Jr in *Philosophic Words: a Study of Style and Meaning in the 'Rambler' and 'Dictionary' of Samuel Johnson* (New Haven: Yale University Press, 1948).

2.   On the controversy between Leibniz and the Newtonians over the interstitial vacuum, see Arnold Thackray, 'Matter in a Nutshell': Newton's *Opticks* and Eighteenth-Century Chemistry', *Ambix*, 15 (1968) 42; and *Atoms and Powers: an Essay on Newtonian Matter-Theory and the Development of Chemistry* (Cambridge, Mass: Harvard University. Press, 1970). I am indebted to Thackray's article and book for my knowledge of this controversy.

3.   The text of Newton's argument deserves to be quoted in full:

> Now if we conceive these Particles of Bodies to be so disposed amongst themselves, that the Intervals or empty Spaces between them may be equal in magnitude to them all; and that these Particles may be composed of other Particles much smaller, which have as much empty Space between them as equals all the Magnitudes of these smaller Particles: And that in like manners these smaller Particles are again composed of others much smaller, all which together are equal to all the Pores or empty Spaces between them; and so on perpetually till you come to solid Particles, such as have no Pores or empty Spaces between them; And if in any gross Body there be, for instance, three such degrees of Particles, the least of which are solid; this Body will have seven times more Pores than solid Parts. It there be five degrees, the Body will have one and thirty times more Pores than solid Parts. If six degrees, the Body will have sixty and three times more Pores than solid Parts. And so on perpetually. And there are other ways of conceiving how Bodies may be exceeding porous. But what is really their inward Frame is not yet known to us. (*Opticks: or, A Treatise of the Reflections, Refractions, Inflections, and Colours of Light*, 4th ed (London, 1730) pp. 268–9).

4.   Sir Henry Pemberton, *A View of Sir Isaac Newton's Philosophy* (Dublin, 1728) p. 3; Colin Maclaurin, *An Account of Sir Isaac Newton's Philosophical Discoveries* (London, 1748) p. 77.

5.   George Cheyne, *An Essay on Regimen* (London, 1757) p. 93; Isaac Watts, *Philosophical Essays on Various Subjects* (London, 1742) p. 2; Isaac Watts, *The Improvement of the Mind* (London, 1789) p. 379.

6. Quoted from Thackray, 'Matter in a Nutshell', pp. 46–7; *Elements of Chemistry* (London, 1735) I, 46–7, 230. On the difficulties Boerhaave's belief in the heterogeneity of matter led him into as a chemist, see Robert E. Schofield, *Mechanism and Materialism: British Natural Philosophy in an Age of Reason* (Princeton University Press, 1970) pp. 150–1.

7. Arthur O. Lovejoy. *The Great Chain of Being: a Study of the History of an Idea* (New York: Harper & Brothers, 1960) pp. 251–2. Lovejoy makes no reference to Newton or his followers in his text.

8. *The Elements of Sir Isaac Newton's Philosphy*, trans. John Hanna (London, 1738) p. 167.

9. *Poem on the Lisbon Disaster*, ft. 1, in *The Portable Voltaire*, ed. Ben Ray Redman (New York: Viking Press, 1949) pp. 563–4.

10. *The Philosophical Dictionary*, ed. and trans. Peter Gay (New York: Harcourt Brace [1962]) p. 163.

11. *Thraliana*, ed. Katherine C. Balderston (2 vols, Oxford: Clarendon Press, 1951) I, 179. See also I, 254.

12. For an excellent discussion of Johnson's critique of Jenyns's essay, see Robert Eberwein, 'Samuel Johnson, George Cheyne, and the Cone of Being', *Journal of the History of Ideas*, 36 (1975) 153–58. I disagree, however, with Eberwein's contention that Johnson's substitution of George Cheyne's 'cone of being' for Jenyns's 'chain of nature' makes it easier for Johnson to argue that there can be infinite vacuities between degrees in the hierarchy (pp. 156–7). Both concepts, in my opinion, are equally susceptible to the kind of argument Johnson advances. Other discussions of Johnson's critique of Jenyns's *Inquiry* include Stuart Gerry Brown, 'Dr. Johnson and the Old Order', Samuel Johnson: *A Collection of Critical Essays*, ed. Donald J. Greene (Englewood Cliffs, N. J.: Prentice Hall, 1965) pp. 158–71; Lester Goodman, 'Samuel Johnson's *Review* of Soame Jenyns's *A Free Enquiry into the Nature and Origin of Evil*: a Reexamination', *New Rambler*, 4 (1968) 19–23; and Richard B. Schwartz, *Samuel Johnson and the Problem of Evil* (Madison: University of Wisconsin Press, 1975).

13. Arieh Sachs, *Passionate Intelligence: Imagination and Reason in the Work of Samuel Johnson* (Baltimore: Johns Hopkins Press, 1967) p. 21.

14. *An Essay on Man*, I, 244–6, in *The Poems of Alexander Pope*, ed. John Butt (New Haven: Yale University Press, 1963) p. 513. In *The Great Chain of Being*, Lovejoy recognizes that the notion of man as middle link between animal and angel reflected a duality that is at odds with the principles of hierarchy and continuity (pp. 80, 127–9, 198).

15 For a fascinating analysis of the emotional undercurrents of Johnson's critique of Jenyns's argument, see William Holtz, 'Samuel Johnson and the Abominable Fancy', *Cithara*, 18 (1979) 29–47.

16. Richard S. Westfall draws attention to the contradictory conclusion Pascal's ambivalence toward the vacuum led him in his experiments on the Toricellian vacuum:

> In his early work on the Torricellian vacuum, Pascal drew from the experiments a conclusion which appears surprising, not to say dumbfounding, to the 20th century reader. The conclusion was that nature abhors a vacuum. A second conclusion substantially modified the first, however – to wit, that nature's abhorrence of a vacuum is finite and measured by the weight of a unit column of

mercury twenty-nine inches high. When a greater force is applied, a vacuum (or at least a space devoid of tangible matter) can be created. What appears as a compromise in fact demanded more than the Aristotelian philosophy could concede because it admitted the possibility of a vacuum under certain conditions. (*The Construction of Modern Science: Mechanism and Mechanics* (New York: Wiley & Sons, 1971) p. 47)

On the intellectual history of the *horror vacui*, see Edward Grant, *Much ado About Nothing: Theories of Space and Vacuum from the Middle Ages to the Scientific Revolution* (Cambridge University Press, 1981) pp. 67−102. For an illuminating discussion of the many parallels that exist between Johnson and Pascal, see Chester Chapin, 'Johnson and Pascal' in *English Writers of the Eighteenth Century*, ed. John H. Middendorf (New York: Columbia University Press, 1971) pp. 3−16.

CHAPTER 2:   JOHNSON'S MORAL PSYCHOLOGY

1. Joseph Wood Krutch, *Samuel Johnson* (New York: Harcourt, Brace & World, 1963; first published in 1944) p. 164.
2. The notion that good and evil, virtue and vice, are inextricably related to one another and that we cannot attain one without involving ourselves in the other is a commonplace in Johnson's moral writings; see, e.g., *Rambler*, I, 76; *Rambler*, II, 34. This commonplace may only acquire real significance, however, when it is seen as an alternative to the traditional polarity between the two terms.
3. William K. Wimsatt, Jr, *Philosophic Words*, p. 101. It should be noted that Wimsatt refers to Johnson's scientific preoccupations in terms of what he calls 'the language of mechanical philosophy', but this phrase is probably still too broad to define the precise nature of Johnson's philosophical orientation.
4. *Meditations on First Philosophy*, in *The Essential Descartes*, ed. Margaret D. Wilson (New York: New American Library, 1969) p. 174. It is possible to see Pope's extension of the doctrine of the plenum to soul as well as body in the contention in *An Essay on Man* that the 'stupendous whole. . . . Breathes in our soul, informs our mortal part,/As full, as perfect, in a hair as heart' (I, ll. 267, 276−7).
5. *Meditation V* in *The Complete Poetry and Selected Prose of John Donne*, ed. Charles M. Coffin (New York: Modern Library, 1952) p. 420. For a discussion of the influence of the principle of plenitude on the moral psychology of Thomas Traherne, see Carol L. Marks, 'Thomas Traherne and Cambridge Platonism', *Publications of the Modern Language Association*, 81 (1966) 530−3.
6. *Philosophical Principles of Religion. Natural and Revealed* (2nd rev. ed, originally published in 1715) II, 124−5. For a brief discussion of this theme in Cheyne, see Hélène Metzger, *Attraction Universelle et Religion Naturelle ches quelques Commentateurs Anglaise de Newton* (Paris: Hermann, 1938) pp. 148−50. On the Newtonian context of Cheyne's *Philosophical Principles*, see Anita Guerini, 'James Keill, George Cheyne and Newtonian Physiology: 1690−1740', *Journal of the History of Biology*, 18 (1985) 260−6.
7. *Philosophical Principles*, II, 76−7.
8. The close link between Johnson's public writings and his private experience has been fruitfully explored in a number of recent studies. See, especially, W. B. C.

Watkins, *Perilous Balance: The Tragic Genius of Swift, Johnson and Sterne* (Cambridge: Walker de Berry, 1960; first published 1939) pp. 58–70; Walter Jackson Bate, *The Achievement of Samuel Johnson* (New York: Oxford University Press, 1961; first published in 1955) pp. 63–91; and Arieh Sachs, *Passionate Intelligence*, pp. 1–12.

9. On the importance of the hunger of the imagination in Johnson's moral vision, see, especially, Bate, pp. 63–71, and Sachs, pp. 3–7. What Johnson has done is to recast the notion that an unquenchable desire for happiness has been implanted in man into psychological terms, shifting it from the soul to a specific power or faculty of the mind. In the process, however, Johnson may have deprived it of its teleological orientation. In other words, there may no longer be any privileged ground of desire from which man could naturally be expected to seek eternal good.

10. Watkins, *Perilous Balance*, pp. 49–51. See also Maurice J. Quinlan, 'The Reaction to Dr Johnson's *Prayers and Meditations*', *Journal of English and Germanic Philology*, 52 (1953) 125–9.

11. Bate, *Achievement*, p. 64.

12. David Hume, *A Treatise of Human Nature: Book I: Of the Understanding*, ed. D. G. C. Macnabb (Cleveland: World Publishing Co., 1962) p. 81. See, also, John Locke, *An Essay Concerning Human Understanding*, II, XIV. On the possible influence of Locke's conception of temporal succession on Johnson's moral writings, see Phyllis Gaba, 'A Succession of Amusements': The Moralization in *Rasselas* of Locke's Account of Time', *Eighteenth-Century Studies*, 10 (1977) 451–63. Raba's argument would be clearer, however, if she had not confused succession with continuity (i.e. succession without 'felt gaps') in her discussion of *Rasselas*.

13. On this theme, see Bate, pp. 71–3 and Chapin, 'Johnson and Pascal', pp. 10–11.

14. Chapin, 'Johnson and Pascal', pp. 8–9.

15. It is thus perhaps because of acoustic interference that Johnson makes the astronomer declare, 'the winds alone, of all the elemental powers, have refused my authority' (*Rasselas*, 46, 136). Jacques Derrida adopts a somewhat similar attitude to the interior voice in *Speech and Phenomena and Other Essays on Husserl's Theory of Signs* (Evanston, Ill: Northwestern University Press, 1973) pp. 75–80; and 'Qual Quelle: Valery's Sources' in *Margins of Philosophy* (University of Chicago Press, 1982) pp. 286–90.

CHAPTER 3:   DESIRE, EMULATION . . . IN *IRENE*

1. Alkon, *Samuel Johnson and Moral Discipline*, pp. 9, 14.

2. For the view that Johnson regards morality in consequentialist terms, see Robert Voitle, *Samuel Johnson, The Moralist*, pp. 127–30, and Alkon, pp. 13–14.

3. Lovejoy, pp. 61–6.

4. *Politics*, V in *The Basic Works of Aristotle*, ed. Richard McKeon (New York: Random House, 1941) pp. 1132–3.

5. See, e.g. Voitle, *Johnson, The Moralist*, p. 105.

6. This denial of an original authority brings Johnson into substantial agreement with those rational, secular theorists of the seventeenth- and eighteenth-centuries (Hobbes, Locke, Rousseau) who defined the origin of society as a product of artifice or contrivance rather than natural instinct. On Johnson's affinities with these philosophers, see Greene, *The Politics of Samuel Johnson*, p. 195. Johnson's ideas

about the origin and development of political societies can be found in his contributions to Sir Robert Chamber's Vinerian lectures. See E. L. McAdam, Jr, *Doctor Johnson and the English Law* (Syracuse University Press, 1951) pp. 81–120. Thomas M. Curley in 'Johnson's Secret Collaboration' in *The Unknown Samuel Johnson*, eds John J. Burke and Donald Kay (Madison: University of Wisconsin Press, 1983) pp. 95–6, accepts McAdam's conclusion that Johnson may have influenced Chambers's lectures but believes that McAdams's assumption that Johnson dictated random passages may be misleading. In this study, the only passages cited from the Vinerian lectures are those that are consistent with the point of view expressed elsewhere in Johnson's writings.

7. On this point, Johnson writes in his contribution to Chambers's Vinerian lectures, 'every man naturally loves dominion and favors that scheme by which his power is advanced' (McAdams, p. 117).

8. Boswell's *Life of Johnson* (London: Oxford University Press, 1952; first published in 1904) pp. 744–5.

9. Ibid., p. 912.

10. On this tradition, see Eugene Waith, *The Herculean Hero in Marlowe, Chapman, Shakespeare, and Dryden* (New York: Columbia University Press: London: Chatto & Windus, 1962); and my *'All for Love* and the Heroic Ideal', *Genre*, XVI (1983) 57–74. The approach of this chapter as well as its title were suggested by Serge Doubrovsky's *Corneille et la dialectique du héros* (Paris: Gallimard, 1963).

11. For a succinct account of Johnson's difficulties in getting *Irene* produced, see the Introduction to *Irene* in *Poems*, pp. 109–10.

12. The argument that the play is undramatic has been advanced by several critics. Among them, Bertrand Bronson, *Johnson Agonistes* (Berkeley: University of California Press,1965), finds an 'untheatrical quality of imagination' in *Irene* (p. 123); Donald Greene, *The Politics of Samuel Johnson*, describes the play as 'not coherently organized or sharply pointed' (p. 79); and, above all, Leopold Damrosch, Jr, *Samuel Johnson and the Tragic Sense* (Princeton University Press, 1972) holds that the play is bookish, boring and untragic (pp. 109–38). *Irene* has been defended by Marshall Waingrow, 'The Mighty Moral of *Irene*' in *From Sensibility to Romanticism, Essays Presented to Frederick A. Pottle* (London: Oxford University Press, 1970; first published 1965), who praises Johnson for his portrayal of Cali and Irene as complex and mixed characters, pp. 79–92; and by Phillip Clayton, 'Samuel Johnson's *Irene*: an Elaborate Curiosity', *Tennessee Studies in Literature*, 19 (1974) 121–35, who insists that while the diction of the play is flawed, it conforms to the dramatic unities.

13. Greene, pp. 75–80, cites Tacitus and Locke as possible sources for this passage. To these, one might also add Aristotle's *Politics* and even passages in Pope's *An Essay on Man*.

14. I strongly disagree with critics who, like Bronson, find that Irene's incapacity to love makes her uninteresting (p. 130), or those who, like Damrosch, find the motives behind her ambition and apostasy base and therefore dramatically flawed (p. 128).

15. Though Aspasia displays throughout *Irene* the courage that Irene lacks, her susceptibility to fear in a moment of crisis (V. v. 10–15) puts her argument on the cultural basis of feminine terrors in question.

16. Damrosch, p. 114, holds that Johnson 'virtually ignores the contrast between the two great civilizations'. In a sense this is true (see below), but only because of the

debacle that has corroded the civic humanist ideal of the Grecian camp.

17. Roughly, I agree with Donald Greene, pp. 74–5, that any 'historical parallel' to the political situation in Walpole's England of the 1730s would have been vague rather than precise. It is uncertain whether all societies based on the civil humanist model are to be seen as susceptible to the kind of degeneration depicted in *Irene*, but there is no question that this degeneration is presented as if it were already inscribed within the genesis of that model (as a dangerous but natural tendency) rather than being superimposed on the model from without.

18. On the theatrical irrelevance of this scene, see Bronson, pp. 114–15. The ambiguity inherent in Demetrius's 'patriotic virtue' is further explored in a late scene in which he resorts to violence (V. v. 47–52) in an attempt to compel Irene to abandon her apostasy and the Turkish cause.

19. On the significance of the theme of means and ends in *Irene*, see Waingrow, pp. 83–5.

CHAPTER 4: THE DECLINE OF THE HEROIC

1. Greene, *The Politics of Samuel Johnson*, pp. 90–1. Critics have often felt uncomfortable with *London*, their objections falling into several broad categories: the poem is a mere exercise, an insincere bid for success, its satire a collection of opposition commonplaces, exaggerated in its attack or blurred in its focus. Behind these objections, one can detect a still deeper reservation, the fear that Thales's tirade – which is marred by prejudice, self-pity, vague hysteria, and hollow posturing – is a reflection of Johnson's own resentments and prejudices as a young man. Among critics who have made such objections are Krutch, *Samuel Johnson*, p. 64; W. J. Bate, *Samuel Johnson* (New York: Harcourt, Brace, 1977) pp. 173–4; T. S. Eliot, 'Johnson as Critic and Poet', in *On Poetry and Poets* (London: Faber & Faber, 1957) p. 205; Paul Fussell, *Samuel Johnson and the Life of Writing* (New York: Harcourt, Brace, 1971) p. 19; and Howard Weinbrot, *The Formal Strain: Studies in Augustan Imitation and Satire* (University of Chicago Press, 1957) p. 179. Critics have generally not attempted to link the political satire and satiric *persona* to an over-arching vision of man and society. Pope evidently admired the poem when it appeared and said that its anonymous author would soon be *deterré* (Boswell, *Life of Johnson*, p. 92).

2. D. V. Boyd, 'Vanity and Vacuity: a reading of Johnson's Verse Satires', *English Literary History*, 39 (1972) p. 395; and T. K. Wharton, *Samuel Johnson and the Theme of Hope* (New York: St Martin's Press, 1984) pp. 35–9. I agree with their approach to Thales, though I strongly disagree with their conclusion that Johnson's satire is compromised. A discussion of Juvenalian satire that emphasizes the inescapable involvement of the satiric *persona* in the vices he is attacking can be found in Alvin Kernan's *The Cankered Muse: Satire of the English Renaissance* (New Haven: Yale University Press, 1959) pp. 14–30. For the view that Johnson's portrayal of Thales is unironic, see Howard Weinbrot, *The Formal Strain*, p. 180, and 'Johnson's *London* and Juvenal's Third Satire: The Country as "Ironic Norm"', *Modern Philology*, 73 (1976) S56–S65.

3. On this point, see Weinbrot, *The Formal Strain*, p. 185. For a sensitive discussion of the way Johnson heightens the irony in this part of his adaptation of Juvenal's third satire, see Edward A. and Lillian D. Bloom, 'Johnson's *London* and the Tools of

Scholarship', *Huntington Library Quarterly*, 34 (1971) 133–9.

4. The transition from *London* to *The Vanity of Human Wishes* has usually been formulated in general terms. See, for example, Krutch, p. 65. For an essay that relates the transition to large stylistic changes that were taking place in English poetry during the 1740s, see John E. Sitter, 'To *The Vanity of Human Wishes* through the 1740s', *Studies in Philology*, 74 (1977) 445–64.

5. In an interesting article entitled 'The Roles of Swift and Marlborough in *The Vanity of Human Wishes*', *Modern Philology*, 73 (1975) 280–3, Colin J. Horne draws attention to the way in which Johnson makes use of Swift's *Examiner* papers and *Conduct of the Allies* in his general satire on the warrior–hero, while at the same time refraining from any direct reference to Marlborough in this section.

6. The effect on readers of the calculated ambiguity in Johnson's irony can be seen in the polarization of contemporary opinion into two camps, a majority of critics arguing that the poem is tragic or heroic in perspective, a minority holding that it is ironic or satiric. The first group includes Ian Jack, *Augustan Satire: Imitation and Idiom in English Poetry 1660–1750* (New York: Oxford University Press, 1952) p. 135; Henry Gifford, '*The Vanity of Human Wishes*', *Review of English Studies*, n.s. 6 (1955) p. 160; Mary Lascelles, 'Johnson and Juvenal' in *New Light on Dr. Johnson: Essays on the Occasion of His 250th Birthday*, ed. Frederick W. Hilles (New Haven: Yale University Press, 1955) p. 55; and, more recently, Michael M. Cohen, 'Johnson's Tragedy of Human Wishes', *English Studies*, 63 (1982) 410–17. Among the second group are Howard Weinbrot, *The Formal Strain*, pp. 203, 206; Donald Greene, *Samuel Johnson* (Boston: Twayne, 1970) p. 57; and William F. Kniskern, 'Satire and the "Tragic Quartet" in *The Vanity of Human Wishes*', *Studies in English Literature*, 25 (1985) 633–50. The argument here asumes that while the ironic is clearly predominant over the tragic or heroic in Johnson's portraits, it is not allowed to crystallize into an affirmation of anti-heroic or anti-tragic ways of life. In Johnson's twlight zone of universal vanity, the status of both heroic and anti-heroic states of mind is called into question.

7. To a certain extent, this pattern, as many critics have noted, resembles the medieval *de casibus* theme, exemplified, for example, in Chaucer's *The Monk's Tale*. What may distinguish Johnson's rendering of this theme from the conventional treatment is that the fall is attributed not merely to an external agency, like fortune or Providence, but also to an internal contradiction within the *libido dominendi* itself.

8. In this view of the theatricality of Charles's campaigns, I am indebted to Lionel Gossman's 'Voltaire's Charles XII: History into Art', *Studies in Voltaire and the 18th Century*, 25 (1963) 691–6.

9. My discussion here follows Macdonald Emslie, 'Johnson's Satires and "The Proper Wit of Poetry"', *Cambridge Journal*, VII (1954) 347–60. I am indebted to his pioneering essay for the initial premise of this section. See, also, the perceptive discussion of the theatrical metaphor in Thomas Jemielity's '*The Vanity of Human Wishes*: Satire Foiled or Achieved', *Essays in Literature*, 11 (1984) 36–8.

10. Bate, *The Achievement of Samuel Johnson*, pp. 21–2. Many readers have found this passage unconvincing for different reasons. Among them are Damrosch, p. 152; and Charles E. Pierce, Jr, 'The Conflict of Faith and Fear in Johnson's Moral Writings', *Eighteenth-Century Studies*, 15 (1982) pp. 321–3. Others have found it moving and essentially satisfying in its Christian orthodoxy. See, e.g. Greene, *Samuel Johnson*, pp. 121–7; Alkon, *Johnson and Moral Discipline*, pp. 197–8; Mary Lascelles, 'Johnson and Juvenal', p. 54; and Elizabeth Macandrew, 'Life in a Maze –

Johnson's Use of Chiasmus in *The Vanity of Human Wishes'*, *Studies in Eighteenth-Century Culture*, 9 (1979) pp. 519–20. A third group, to which this interpretation belongs, finds the conclusion moving, consistent with the remainder of the poem, and yet quite limited in the consolation it offers the believer. See. e.g. Boyd, 'Vanity and Vacuity', p. 403; and George Amis, 'The Style of *The Vanity of Human Wishes'*, *Modern Language Quarterly*, 35 (1974) 16–29.

11.   G. W. F. Hegel, *Phenomenology of Spirit*, trans. A.V. Miller (Oxford University Press, 1977) pp. 126–38. The imagery of Hegel's analysis makes it clear that he regards the unhappy consciousness as an exemplification of an orthodox medieval conception of the petitioner's relation to God.

12.   This view is consistent with Johnson's adherence throughout most of his life to what Maurice J. Quinlan has described as the doctrine of exemplary rather than propitiatory atonement (*Samuel Johnson: A Layman's Religion* [Madison: University of Wisconsin Press, 1964] pp. 46–65. According to Quinland, Johnson's thinking about atonement may have undergone a shift in the last two years of his life (p. 197).

CHAPTER 5: 'THE MAZE OF VARIATION'

1.   Willard Van Orman Quine, *From a Logical Point of View* (Cambridge, Mass.: Harvard University Press, 1953) p. 59.

2.   For the traditional view, see Albert C. Baugh and Thomas Cable, *A History of the English Language* (3rd ed.: Englewood Cliffs, N.J.: Prentice Hall, 1978) pp. 270–2. De Witt T. Starnes and Gertrude E. Noyes, *The English Dictionary from Cawdrey to Johnson, 1604–1755* (Chapel Hill: University of North Carolina Press, 1946) maintain that Johnson was recognized by his contemporaries as the inheritor of all schemes for fixing the language from the time of the Academie Francaise to the present (p. 273n. 23).

3.   See W. Scott Elledge, 'The Naked Science of Language, 1747–1786' in *Studies in Criticism and Aesthetics, Essays in Honor of Samuel Holt Monk*, eds Howard Anderson and John S. Shea (University of Minnesota Press, 1967) 266–95; and Leo Braudy, 'Lexicography and Biography in the *Preface* to Johnson's *Dictionary'*, *Studies in English Literature*, 10 (1970) 551–6. For a view of Johnson as 'the acknowledged source' of the drastically reduced aims of late eighteenth-century linguists, see Murray Cohen, *Sensible Words: Linguistic Practice in England, 1640–1785* (Baltimore: Johns Hopkins University Press, 1977) pp. 89–94. An alternative to the standard account of Baugh and Cable can be found in Thomas Pyle's *The Origins and Development of the English Language* (New York: Harcourt, Brace, 1971) pp. 224–5.

4.   For an interesting analysis of some of the theoretical aspects of Johnson's *Plan* and *Preface*, see C. H. Knoblauch's 'Coherence Betrayed: Samuel Johnson and the "Prose of the World"', *Boundary 2*, 7 (1979) 235–60. This article discusses some of the issues raised in this chapter but not with any thoroughness.

5.   On this point, see John Barrell, *English Literature in History, 1730–1780: an Equal Wide Survey* (New York: St Martin's Press, 1983) p. 152. While I agree with Barrell

that Johnson shared Burke's hostility to innovation, I strongly disagree with the conclusion that Johnson's conservatism stems from the same vision of historical change as that of Burke.

6.   Francis Bacon, *Essays, Advancement of Learning, New Atlantis, and Other Pieces*, ed. Richard Foster Jones (New York: Odyssey Press, 1937) pp. 287–8.

7.   Quoted by Stephen K. Land, *From Signs to Propositions: The Concept of Form in Eighteenth-Century Semantic Theory* (London: Longmans, 1974) p. 20; from *Introduction to Languages* (London: 1738) III, 79.

8.   The abandonment in the *Preface* of the systems of classification proposed in the *Plan* has been noted by Knoblauch, p. 244.

9.   This is the argument of Dorothy Bilik, in 'Johnson Defines an Audience for the Dictionary', *New Rambler*, 17 (1976) 45–9. Knoblauch, on the other hand, believes that the *Preface* 'effectively sabotages the achievement of the *Dictionary* and mocks explicitly, just as the flawed text of the work implicitly mocks, the inadequacy of the author' (p. 246).

10.   Hans Aarsleff, *From Locke to Saussure: Essays on the Study of Language and Intellectual History* (London: Athlone Press, 1982) p. 163.

11.   Barrell, p. 145.

12.   Thomas B. Gilmore, Jr in 'Johnson's Attitudes toward French Influence on the English Language', *Modern Philology*, 78 (1981) 243–60, draws attention to the 'ambiguity' inherent in these attitudes in both the *Preface* and *Dictionary*.

13.   Ferdinand de Saussure, *Course in General Linguistics*, trans. Wade Baskin (New York: McGraw-Hill, 1959) p. 9.

14.   See Howard D. Weinbrot, 'Samuel Johnson's *Plan* and *Preface* to the *Dictionary*: The Growth of a Lexicographer's Mind' in *New Aspects of Lexicography*, ed. Howard D. Weinbrot (Carbondale: Southern Illinois University Press, 1972) pp. 73–94.

15.   For a survey that locates Johnson in the tradition of eighteenth-century lexicography, see James H. Sleded and Gwin J. Kolb, *Doctor Johnson's Dictionary: Essays in the Biography of a Book* (Chicago: University of Chicago Press, 1955). Paul J. Korshin shows that this tradition should be extended back to the Renaissance in 'Johnson and the Renaissance Dictionary', *Journal of the History of Ideas*, 35 (1974) 300–12.

16.   Saussure, p. 133.

17.   Jacques Derrida, 'Structure, Sign, and Play in the Discourse of the Human Sciences' in *Writing and Difference*, trans. Alan Bass (University of Chicago Press, 1978) p. 289.

CHAPTER 6: ART AND NATURE IN *RASSELAS*

1.   My conception of an apparently unending dialectical conflict between opposites in *Rasselas* has been anticipated by Earl R. Wasserman's point in 'Johnson's *Rasselas*: Implicit Contexts', *Journal of English and Germanic Philology*, 74 (1975) 9, 11, that while 'everything is bipolar, not multiple' in Johnson's tale, there is 'no clear choice . . . but only an endless, directionless oscillation between opposites, neither of which is sufficient or stable'. See also Irvin Ehrenpreis's observation concerning Johnson's method of 'offering a choice of alternatives and undercutting both' in

'*Rasselas* and Some Meanings of "Structure" in Literary Criticism', *Novel*, 14 (1981) 113.

2. Thomas R. Preston, in 'The Biblical Context of Johnson's *Rasselas*', *Publications of the Modern Language Association*, 84 (1969) 275, points to the resemblance between the Happy Valley and the Preacher's Garden in Symon Patrick's *A Paraphrase upon the Book of Ecclesiastes* (1685). To grasp what is unusual about *Rasselas*, however, one must take account of differences as well as resemblances. In striking contrast to the Happy Valley, Patrick's Garden is not organized in terms of a system of classification and exclusion; we learn merely that 'besides other delights', it included 'lovely Shades and Coverts for all Sorts of Beasts', Symon Patrick, *A Paraphrase upon the Book of Ecclesiastes* (London: Royston, 1685) p. 30.

3. The idea that there is an underlying continuity between life inside and outside the Happy Valley has been noted by several critics, among them, Ehrenpreis, p. 106, and Alvin Whitley, 'The Comedy of *Rasselas*', *English Literary History*, 23 (1956) 51.

4. Though Patrick, as Preston notes ('Biblical Contexts', p. 276), advances a similar argument about the inevitability of oppression in even the best-ruled kingdoms, Johnson's argument is presented not as the simple frustration of a desire for security, but as the dissolution of a dialectic of inside and outside.

5. Critics differ as to the significance of the exchanges of the central characters in *Rasselas*. At one extreme, Bertrand H. Bronson, in 'Postscript on *Rasselas*' in *Rasselas, Poems, and Selected Prose* (New York: Holt, Rinehart & Winston, 1952), xvi, contends that the characters are voices in a philosophic dialogue. At the other extreme, Carol J. Sklenicka, in 'Samuel Johnson and the Fiction of Activity', *South Atlantic Quarterly*, 78 (1979) 214–23, and Catherine Neale Parke, in 'Imlac and Autobiography', *Studies in Eighteenth-Century Culture*, 6 (1977) 183–98, contend that the role of conversation in *Rasselas* is therapeutic, protecting the characters from the world's miseries and enabling them to achieve at least a degree of felicity. By contrast, Frederick M. Keener, in *The Chain of Becoming* (New York: Columbia University Press, 1983), tries to show 'the main character's disagreement, resentment, and estrangement from each other' (p. 218).

6. Howard D. Weinbrot, in 'The Reader, the General and the Particular: Johnson and Imlac in Chapter Ten of *Rasselas*', *Eighteenth-Century Studies*, 5 (1971) 86–90, carefully traces the way in which the terms 'species' and 'genus' are opposed in Johnson's critical vocabulary. What my interpretation emphasizes is the absence of any theory of analogy or correspondence that might serve as a bridge between the two terms.

7. On the demise of patriarchalism in the English political thought of the early eighteenth century, see Isaac Kramnick, *Bolingbroke and His Circle: the Politics of Nostalgia in the Age of Walpole* (Cambridge, Mass.: Harvard University Press, 1968) *passim*; and Gordon J. Schochet, *Patriarchalism in Political Thought* (New York: Basic Books, 1975) pp. 192–4.

8. I disagree with critics who view the conclusion as the culmination and perpetuation of a movement which is closed, circular, and endless. See, for example, Whitley, p. 69; Wasserman, p. 25; and Emrys Jones, 'The Artistic Form of *Rasselas*', *Review of English Studies*, N. S. 18 (1967) 400. The tenor of my interpretation of *Rasselas* inclines to the view that the conclusion should be seen in terms of rupture rather than continuity, succession rather than incessant motion, unpredictable reversal rather than monotonous circular repetition.

9. For an interpretation that views Imlac as a normative figure, see the fine essay by

Agostino Lombardo, 'The Importance of Imlac', in *Bicentenary Studies on Rasselas*, Supplement to *Cairo Studies in English* (Cairo: n. p., 1959) pp. 31–9. See, in addition, the importance attached to Imlac's wisdom in William Kenny's 'Rasselas and the Theme of Diversification', *Philological Quarterly*, 38 (1959) 84–9; and Frederick M. Keener's *The Chain of Becoming*, p. 237.

10. On the eighteenth-century intellectual background of the arguments Imlac advances in his debate with the astronomer on the immortality of the soul, see Carey McIntosh, *The Choice of Life: Samuel Johnson and the World of Fiction* (New Haven: Yale University Press, 1973) pp. 202–4; and, more fully, Gwin J. Kolb, 'The Intellectual Background of the Discourse on the Soul in *Rasselas*', *Philological Quarterly*, 54 (1975) 357–69. For an interesting discussion of the role of Christianity in *Rasselas*, see Nicholas Joost's 'Whispers of Fancy; or, the Meaning of *Rasselas*', *Modern Age*, 1 (1957) 168–73.

CHAPTER 7: *A JOURNEY TO THE WESTERN ISLANDS*

1. For a discussion of the contemporary recognition of the originality of Johnson's *Journey*, see Clarence Tracy, 'Johnson's *Journey to the Western Islands of Scotland*: A Reconsideration', *Studies in Voltaire and the Eighteenth Century*, 58 (1967) 1596–8. A. J. Youngson, in *Beyond the Highland Line, Three Journals of Travel in Eighteenth-Century Scotland* (London: Collins, 1974) p. 11, describes Johnson's *Journey* as 'in many ways the best book about Scotland in the eighteenth century'.

2. Johnson's extensive acquaintance with the travel literature of his age has been convincingly demonstrated by Thomas Jemielity, in 'Doctor Johnson and the Uses of Travel', *Philological Quarterly*, 51 (1972) 450–2. Jemielity also provides a useful summary of Johnson's theory of travel in *Idler*, No. 97 (pp. 453–5).

3. Patrick O'Flaherty, 'Johnson in the Hebrides: Philosopher Becalmed', *Studies in Burke and His Time*, 13 (1971) 1986–2001. Rejecting the notion of Johnson as anthropologist, O'Flaherty seems him 'as a class-conscious Tory imposing his own preconceptions upon primitive Scotland' (p. 1999).

4. Patrick Cruttwell, ' "These are not Whigs": 18th Century Attitudes toward the Highlanders', *Essays in Criticism*, 15 (1965) 398, 401; Donald Greene, *Samuel Johnson* (Boston: Twayne, 1970) pp. 169–75. There is some generic confusion as to the more precise definition of Johnson's activity in *A Journey*. Thus, for example, several critics have compared his investigations to that of a 'modern sociologist'. See Jemielity, *ibid*, p. 458; Tracy, p. 1600; and Greene, in 'Johnsonian Critics', *Essays in Criticism*, 10 (1960) p. 479. Johnson differs from modern sociologists, however, in his preoccupation with a culture that is clearly alien and more backward than his own. By contrast, Johnson was obviously reluctant to write about Wales or contemporary France (see, Jemielity, ibid., pp. 457–9). The term 'cultural anthropology' is perhaps the most satisfactory description of Johnson's work. R. N. Kaul in '*A Journey to the Western Islands* Reconsidered', *Essays in Criticism*, 13 (1963) 343, describes the *Journey* as the work of a 'social anthropologist'. Although a part of social anthropology overlaps with cultural

anthropology, Johnson seems much more interested in material culture than in contemporary social structures and institutions. On the difference between the two disciplines, see Claude Lévi-Strauss, 'The Place of Anthropology in the Social Sciences', in *Structural Anthropology*, trans. Claire Jacobson and Brooke Grundfest Schoepf (New York: Basic Books, 1963) pp. 356—9.

5. Mary Lascelles, in 'Some Reflections on Johnson's Hebridean Journey', *New Rambler*, 1 (1961), p. 8, praises Johnson's 'grasp of the whole' of Highland life. See, also, Thomas Jemielity's discussion of Johnson's 'generalized presentation of an entire society' in '"More in Notions than in Facts": Samuel Johnson's *Journey to the Western Islands*', *Dalhousie Review*, 49 (1969) 322—4.

6. See, especially, the sections entitled the Highlands (43—9), Coriatachan in Sky (53—8), Raasay (58—71), Ostig in Sky (79—120), Castle of Col (124—36), and Inch Kenneth (142—64).

7. E. B. Tylor, *Primitive Culture* (London, 1871) I, i; quoted from Claude Lévi-Strauss, ibid., p. 356.

8. For a survey of the eighteenth-century anthropological texts of Montesquieu, Kames, Millar, and Condorcet, see Sir Edward Evans-Pritchard, *A History of Anthropological Thought* (New York: Basic Books, 1981) pp. 3—40.

9. The best discussion of Johnson's cautious and sceptical handling of evidence can be found in Francis Hart's 'Johnson as Philosophic Traveller: The Perfecting of an Idea', *English Literary History*, 36 (1969) 679—95. See also the valuable discussions of Richard Schwartz, 'Johnson's *Journey*', *Journal of English and Germanic Philology*, 69 (1970) 292—303; Thomas Curley, 'Johnson and the Geographical Revolution: *A Journey to the Western Islands*', *Studies in Burke and His Times*, 17 (1976) 180—96; and Catherine N. Parke, 'Love, Accuracy, and the Power of an Object: Finding the Conclusion in *A Journey to the Western Islands*'., *Biography*, 3 (1980) 115—20.

10. For an interesting discussion of the ahistorical bias to which the functionalist—holistic tendency can lead in modern anthropology, see Clifford Geertz, 'Ritual and Social Change: A Javanese Example', in *The Interpretation of Cultures* (New York: Basic Books, 1973) pp. 142—7. I am indebted to Geertz's discussion in my emphasis on Johnson's attention to the disruptive and psychologically disturbing aspects of Highland life.

11. For a detailed analysis of the *Journey* as a romantic quest narrative, see Eithene Henson, 'Johnson's Quest for "The Fictions of Romantic Chivalry" in Scotland', *Prose Studies*, 7 (1984) 97—128. Thomas R. Preston in 'Homeric Allusion in *A Journey to the Western Islands*', *Eighteenth-Century Studies*, 5 (1972), 545—58, has drawn attention to the Homeric framework of Johnson's *Journey*. See, also, George H. Savage, ' "Roving Among the Hebrides": the Odyssey of Samuel Johnson', *Studies in English Literature*, 17 (1977) 493—501.

12. George Cheyne, *An Essay on Regimen* (London: Browne, 1757) pp. 106—7.

13. On Johnson's use of the ruins as 'a tangible symbol of deeper decay', see John A. Vance, *Samuel Johnson and the Sense of History* (Athens: University of Georgia Press, 1984) pp. 75—80.

14. Thomas J. Jemielity, in ' "Savage Virtues and Barbarous Grandeur": Johnson and Martin in the Highlands', *Cornell Library Journal*, I (1966) 1—12, shows Johnson's disappointment with Martin Martin's *Description of the Western Islands of Scotland* (1703), for Martin's failure to recreate Highland culture when the opportunity to do so still existed.

15. The ambivalence of Johnson's response gave rise to a debate as to whether he

should be viewed as a Tory defender of the old order or an advocate of commerce and manufacture. See especially Jeffrey Hart, 'Johnson's *Journey to the Western Islands*: History as Art', *Essays in Criticism*, 10 (1960) 44–59; and Thomas K. Meier, 'Johnson on Scotland', *Essays in Criticism*, 18 (1968), for the first view; and Donald Greene, 'Johnsonian Critics', *Essays in Criticism*, 10 (1960) 476–80; R. K. Kaul, '*A Journey to the Western Islands* Reconsidered', *Essays in Criticism*, 13 (1963) 341–50; and Arthur Sherbo, 'Johnson's Intent in the *Journey to the Western Islands of Scotland*', *Essays in Criticism*, 16 (1966) 382–97, for the second. A recent, balanced presentation can be found in Jemielity's ' "More in Notions than in Facts" ', pp. 326–8.

16.   Pennant writes in *A Tour in Scotland*, for example, that Perth 'as well as all *Scotland*, dates its prosperity from the year 1745; the government of this part of *Great Britain* having never been settled till a little after that time. The rebellion was a disorder violent in its operation, but salutary in its effects', *Beyond the Highland Line*, p. 127. For a discussion of Johnson's relation to Pennant, see Ralph E. Jenkins, ' "And I travelled after him": Johnson and Pennant in Scotland', *Texas Studies in Language and Literature*, 14 (1972) 445–62.

17.   On the analogy between Johnson's conception of Highland feudalism and the Hobbesian state of nature, see Kaul, p. 344, and Preston, p. 346.

18.   Schochet, *Patriarchalism in Political Thought*, pp. 225–43.

19.   Fernand Braudel, *The Mediterranean and the Mediterranean World in the Age of Phillip II*, trans. Sian Reynolds (2 vols, New York: Harper & Row, 1972) I, 25–52.

20.   See, especially, the stories of the inhabitants of the isle of Egg (*Journey*, 69) and Hugh Macleod (*Journey*, 73–4).

21.   Johnson makes a similar comment in the *Journey*: 'Popery is favourable to ceremony; and among ignorant nations, ceremony is the only preservative of tradition. . . . We therefore who came to hear old traditions, and see antiquated manners, should probably have found them amongst the Papists' (pp. 127–8).

22.   Hayden White, *Metahistory: the Historical Imagination in Nineteenth-Century Europe* (Baltimore: Johns Hopkins University Press, 1973) p. 54. see also pp. 60–2. Johnson's conception of history may resemble what White calls the '*historiography of essential schism*'. According to White, this historiography is characteristic of the seventeenth rather than the eighteenth century and apprehends 'the historical field as a chaos of *contending* forces, among which the historian had to choose and in the service of one or more of which he had to write his history. This was the case with both the confessional historiography . . . and the Ethnographic historiography of the missionaries and *conquistadores*' (p. 65).

23.   Cruttwell, in 'These are not Whigs', finds a 'continual to-and-fro movement' throughout the *Journey*. 'No sooner has Johnson said something sympathetic and admiring than he balances it with something critical' (p. 403). For a sensitive and intricate analysis of Johnson's shifting attitudes toward the Conquest, see John B. Radner, 'The Significance of Johnson's Changing Views of the Hebrides', in *The Unknown Samuel Johnson*, eds John J. Burke, Jr. and Donald Kay (Madison: University of Wisconsin Press, 1983) pp. 148–9 n.

24.   On these characteristics of modern anthropology, see Ernest Gellner's introduction to Evans-Pritchard's *A History of Anthropological Thought*, xix–xxii.

Index

148 *Index*

Swift, Jonathan, 63, 138n

Tacitus, 136n
Tarbet, David W., viii
technology, 92–3, 95–7, 107, 109–10,
    130
Thackray, Arnold, 132–3n
Thrale, Hester Lynch, 5, 133n
time, 17–20
Tracy, Clarence, 142n
Traherne, Thomas, 134n
Tylor, E. B., 104, 143n

'unhappy consciousness', 65

vacuities of existence, vii–viii, 4–6, 9–
    10, 13–15, 28, 84, 95, 129–30
vacuum, vii–viii, 1–5, 10–12, 27, 68, 126
Vance, John A., 143n
Voitle, Robert, 135–6n
Voltaire, 4, 6; *The Elements of Sir Isaac*

Newton's Philosophy, 4, 133n; *Poem on
    the Lisbon Disaster*, 4, 133n;
    Philosophical Dictionary, 4, 133n

Waingrow, Marshall, 136–7n
Waith, Eugene, 136n
Wasserman, Earl R., 140–1n
Watkins, W. B. C., 16, 135n
Watts, Isaac, vii; *Philosophical Essays on
    Various Subjects*, 3, 133n; *The
    Improvement of the Mind*, 133n
Weinbrot, Howard W., viii, 137–8n,
    140–1n
Westfall, Richard, 133–4n
Wharton, T. K., 49, 137n
White, Hayden: *Metahistory*, 122, 144n
Whitley, Alvin, 141n
Wimsatt, William K., Jr., viii, 12, 132n
Wright, John W., viii

Youngson, A. J., 142n